C000293682

Introduction to Microelectromechanical Microwave Systems

Second Edition

For a listing of recent titles in the *Artech House Microelectromechanical Systems (MEMS) Series*, turn to the back of this book.

Introduction to Microelectromechanical Microwave Systems

Second Edition

Héctor J. De Los Santos

Artech House
Boston • London
www.artechhouse.com

Library of Congress Cataloging-in-Publication Data
A catalog record of this book is available at the Library of Congress

ISBN 1-58053-871-1

British Library Cataloguing in Publication Data
A catalog record of this book is available at the British Library of Congress

Cover design by Igor Valdman

© 2004 ARTECH HOUSE, INC.
685 Canton Street
Norwood, MA 02062

All rights reserved. Printed and bound in the United States of America. No part of this book may be reproduced or utilized in any form or by any means, electronic or mechanical, including photocopying, recording, or by any information storage and retrieval system, without permission in writing from the publisher.

All terms mentioned in this book that are known to be trademarks or service marks have been appropriately capitalized. Artech House cannot attest to the accuracy of this information. Use of a term in this book should not be regarded as affecting the validity of any trademark or service mark.

International Standard Book Number: 1-58053-871-1

10 9 8 7 6 5 4 3 2 1

Este libro lo dedico a mis queridos padres
y a mis queridos Violeta, Mara, Hectorcito, y Joseph.

"Y sabemos que a los que aman a Dios todas las cosas les ayudan a bien, esto es, a los que conforme a su propósito son llamados."
Romanos 8:28

Contents

Preface

This book deals with the emerging field of microelectromechanical systems (MEMS) fabrication and device technologies as applied to RF and microwave wireless communications systems. The field is interdisciplinary, in that it integrates the areas of integrated circuit fabrication technology, mechanical and materials science, engineering of microscopic mechanisms, and RF and microwave electronics engineering, to create devices and techniques that have the potential to greatly improve the performance of communications circuits and systems. Accordingly, the book integrates the fundamentals of all these disparate areas into a coherent presentation, providing a unified framework that makes the field accessible to both the specialists within the different areas, as well as to newcomers. *An Introduction to Microelectromechanical Microwave Systems, Second Edition, provides* an update on the development of the field over the last five years, and continues to be a timely resource for the diverse myriad of workers already involved in, or interested in, capturing the potential of RF MEMS to revolutionize both wireless portable, base stations, and satellite communications. The book, which assumes a basic preparation at the B.S. in physics or engineering level, is intended for senior level/beginning graduate students, practicing RF and microwave engineers, and MEMS device researchers.

An Introduction to Microelectromechanical Microwave Systems, Second Edition contains six chapters. Chapter 1 starts by reviewing the origins, impetus, and motivation of the MEMS field, putting into perspective its development up to the present, and describing the fundamentals of the technology in view of its relationship to the well-known integrated circuit fabrication technology. Chapter 2 deals with the fundamental physics of electrostatic, piezoelectric, thermal, and magnetic actuation, which are the principles on which most MEM devices applied in RF and microwave systems operate. This chapter also introduces the rudiments of computer-aided design (CAD) numerical and analytical simulation techniques, which are crucial for the rapid prototyping, development, and commercialization of MEMS. Chapters 3 and 4 describe the physics and circuit models for the fundamental MEM devices, namely, the switch and the resonator, respectively. Particular attention is paid to issues like device engineering, design tradeoffs, and performance limitations, and their impact on RF and microwave circuit applications. Chapter 5 surveys and thoroughly explains RF and microwave MEMS applications. In a systematic way, the operation, fabrication, and performance of the fundamental MEM devices and techniques, as applied to switches, resonators, transmission lines, and lumped passive elements, is discussed, with emphasis on the interplay between fabrication process and microwave performance. Finally, Chapter 6 presents the fundamentals of wireless communications systems and the application of MEMS-based devices and techniques in these systems. Since a large portion of the book's intended readership includes materials, process and device scientists and engineers, who may not be familiar at all with RF and microwave electronics, our presentation follows a semi-intuitive, common-sense approach, complete with the fundamentals and MEMS applications to a variety of important RF and microwave circuits, namely, phase shifters, resonators, filters, and oscillators.

Acknowledgments

The second edition of this book would not have been possible without the assistance of those who made the first edition possible. Therefore, the author continues to gratefully acknowledge the overall assistance and cooperation from the former Hughes Space and Communications Company and Hughes Electronics Corporation during this effort. Special thanks are extended to Drs. Michael J. Delaney and Victor S. Reinhardt for their technical review of each chapter and valuable suggestions, to Dr. Hutan Taghavi for his review of Chapter 2 and valuable suggestions, and to Ms. E. E. Leitereg, Ms. F. Slimmer, Ms. D. Ball, Ms. N. Anderson, Mr. R. Fanucchi, and Mr. M. Sales, whose help was essential in expediting the manuscript clearance process. The help of Ms. Leitereg was crucial to the overall success of the project. The excellent assistance provided by Librarians Ms. M. S. Deeds, Ms. B. A. Hinz, and Ms. M. J. Saylor is gratefully acknowledged. Dr. J. Yao of Rockwell Science Center is acknowledged for most graciously providing original artwork, Prof. C. J. Kim of UCLA is acknowledged for providing some hard-to-find references, and Dr. Yu-Hua Kao of HSC is thanked for proofreading various parts of the manuscript. The author gratefully acknowledges the assistance of Mr. Tim Lee, Boeing Satellite Systems, and Dr. Yu-Hua Kao Lin, who helped him gain timely access to many technical articles. The assistance of the staff at Artech House is gratefully acknowledged, in particular, Mr. Mark Walsh, senior acquisition editor, for facilitating the opportunity to work in this project, and Ms. Barbara Lovenvirth, assistant editor, for her assistance throughout manuscript development, and Jill Stoodley, production editor, for her assistance during production. Finally, the author gratefully acknowledges the encouragement and understanding of his wife, Violeta, throughout the course of both these efforts.

1

Microelectromechanical Systems

1.1 MEMS Origins

It can be said that the field of microelectromechanical systems (MEMS) was originated by Richard P. Feynman in 1959, when he made the observation: "There's plenty of room at the bottom" [1].

Feynman arrived at this conclusion after conducting a special type of search, namely, a search for a new field of scientific inquiry that, like those of attaining low temperatures or attaining high pressures, embodied a vast, virtually boundless, territory. The field, he concluded, would be analogous to these was *miniaturization of systems*. The pervading thesis in his presentation [1] was that while, in and of themselves, the laws of physics posed no limit to miniaturization, it was our ability to make physically small things which indeed limited the degree of such miniaturization. Having acknowledged that, at the time, there were severe technological limitations for making physically small things, Feynman nevertheless embarked on a fascinating visionary journey, in which he explored the opportunities and challenges that would arise from attempting to manipulate and control things on a small scale. In his journey, Feynman began exploring the miniaturization of things that did not require any particular size, like numbers, information storage, and computing [1], and concluded with those that did, namely, machinery [2]. In a recent review of Feynman's prophetic lectures, Senturia [3], most appropriately, codified these areas as miniaturization of information storage, computers, atomic-level manipulation, and machinery [1–3].

In the area of miniaturization of information storage, Feynman considered the prospect of increasing memory capacity by miniaturizing the material volume required to represent an element of information. His approach was based on scaling. Indeed, the smaller the physical volume used to represent an element of information, the greater the number of elements that can be packed in the original volume they occupied before being scaled down, so that as the reduction factor increases, more room (plenty) is created. In a concrete example [1], Feynman estimated that the application of a reduction factor of 25,000 to the size of an ordinary printed letter would yield a material volume equivalent to that occupied by 100 atoms which, consequently, would make it possible to pack 10^{15} bits of information in a *speck of dust*.

In the area of miniaturization of computers [1], Feynman prophesied the need for greatly increasing their computing power. His approach was again based on scaling. Scaling down the size of logic circuits reduces their parasitic capacitance, thus simultaneously increasing their speed and reducing their power consumption, as well as increasing their packing density, Increasing their packing density, in turn, leads to an increased computing capacity. This vision has clearly materialized by way of the computer revolution the world has witnessed. Another outcome of the emphasis on continued miniaturization of circuits has been the fact that quantum effects (i.e., the manisfestation of the quantum mechanical nature of electrons), have come to dominate the behavior of the very small electronic devices needed to make very small computers, thus ushering the field of *nanoelectronics*. Nanoelectronics motivated, and makes use of, the development of a number of atomic-level manipulation techniques (e.g., *molecular beam epitaxy)*, to fabricate nanometer feature size devices.

In the area of miniaturization of machinery [1, 2], Feynman pondered on the problems that could be encountered in the design of small machines, taking into consideration the realms of forces, materials, magnetic behavior, and friction. Regarding the design of mechanical elements, for example, he noted that, for a constant stress level, as the element is downscaled, the role played by its weight and inertia becomes insignificant compared to that played by the material strength. Regarding the structural properties of downscaled materials [2], he noted that their grain structure, and thus inhomogeneity, would come to the fore. This, for example, would make metallic materials undesirable to work with. On the other hand, he observed, the interest would be on materials amorphous in nature (e.g., plastics and glass), since these are homogeneous on a small scale. Regarding the magnetic behavior of downscaled materials [1], he observed that because of their domain-based nature, it would be difficult to maintain their large-size magnetic properties. To illustrate this point, he submitted that a big magnet made of millions of domains, for example, must of necessity contain fewer domains on a small scale. This led him to suggest that, to work at small scales, magnetic-based equipment had to be altogether reengineered. Lastly, regarding the issue of friction, Feynman ventured that scaled-down mechanisms would dispose of the need for lubrication. He reasoned that the additional force available to scaled-down mechanisms would counter friction, and that the small sizes involved would facilitate rapid heat transfer and escape, thus precluding heat dissipation problems. Verifying the certainty of lubrication-free operation has been elusive, since all systems built thus far (e.g., micromotors), have exhibited wear and tear [3]. There is hope for the future, however, as recent theoretical and experimental investigations have yielded the discovery and confirmation of *superlubricity*, a new regime in which friction completely vanishes [4].

Feynman's first concrete example of a micromachine was that of an electric micromotor [2]. This system, according to him, would be constructed by a thin-film deposition process and would be operated by *electrostatic actuation*.

1.2 MEMS Impetus/Motivation

Our ability to make physically small objects received a big impetus with the advent of integrated circuit (IC) fabrication technology in the 1960s. In this technology, photographic and chemical etching techniques are used to print circuits on a substrate (wafer). Since the circuits could be scaled, and still perform the same function, a race ensued to develop ways of printing more and more circuits on a semiconductor wafer. On the economic side this was beneficial, because the greater the number of circuits that could be printed on a given wafer area, the greater the profits. Using the number of devices on a circuit within a wafer as an index of the extent to which integration or miniaturization has progressed, we deduce that the number of devices on a circuit has increased by more than seven orders of magnitude, namely, from less than 10 in the 1960s, to more than 10 million from the 1990s.

The witnessing of such an exemplary success in mass production, as displayed by the IC industry, motivated the pursuit of applying the concepts of integrated electronics manufacturing to mechanics, optics, and fluidics, with the hope of reaping the same astounding improvements in performance and cost effectiveness experienced by the semiconductor industry [5]. When considering the extention of IC fabrication techniques to micromechanical structures, however, one fundamental difference became apparent, namely, whereas the domain of the first is two dimensional, that of the latter is three dimensional. Thus the first step in the realization of small micromechanical structures would to have to await the development of fabrication techniques for sculpting the three spatial dimensions [5].

1.3 MEMS Fabrication Technologies

Our goal in this section is to introduce the basic fabrication technologies used in the realization of three-dimensional microelectromechanical systems. A basic knowledge of conventional integrated circuit fabrication processes is assumed [6], but is briefly reviewed to motivate an appreciation for MEMS fabrication technologies.

1.3.1 Conventional IC Fabrication Process

The conventional IC fabrication process is based on photolithography and chemical etching [6], as shown in Figure 1.1:

1. The wafer is first covered with a thin-film material barrier (typically SiO_2 in silicon technology or silicon nitride in gallium arsenide technology) on which a pattern of holes will be defined.
2. The thin-film material, on the wafer surface, is then coated with a light-sensitive material, called photoresist.

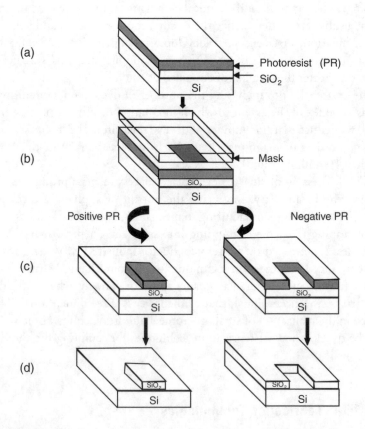

Figure 1.1 Resist and silicon dioxide patterns following photolithography with positive and negative resists. (*After:* [6].)

3. A photomask, a square glass plate with a patterned emulsion or metal film on one side is placed over the wafer and the photoresist is exposed through the mask to high-intensity ultraviolet (UV) light wherever the oxide is to be removed.

4. The photoresist is developed with a process very similar to that used for developing ordinary photographic film. The parts of the photoresist film that were exposed to the UV light, may or may not wash out according to whether it is of positive or negative type.

- If the resist which has been exposed to UV light is washed away, leaving bare SiO_2 in the exposed area, it is positive resist and the mask contains a copy of the pattern, which will remain in the surface of the wafer.

- A negative resist remains on the surface wherever it is exposed.

This is the essence of a conventional IC fabrication process; it permits the transfer of a circuit layout pattern onto a two-dimensional wafer surface.

Conventional processes, however, are not empowered to realize three dimensional structures (i.e., to sculpt three dimensional microelectromechanical devices). This is the realm of two processes devised for this purpose, namely, bulk micromachining and surface micromachining [5].

1.3.2 Bulk Micromachining

Bulk micromachining addresses the creation of mechanical structures in the wafer bulk, as opposed to on its surface. This entails the selective removal of some parts of the wafer/substrate material, as shown in Figure 1.2.

It is considered a mature technology, as it was originally developed for the production of silicon pressure sensors in the late 1950s. Recently [7], however, new techniques aimed at producing three-dimensional micromechanical structures have emerged: surface micromachining, silicon fusion bonding, and a process called the LIGA, a German acronym consisting of the letters LI (RoentgenLIthographie, meaning X ray lithography), G (Galvanik, meaning electrodeposition), and A (Abformung, meaning molding) process. As these technologies are based on the use of photolithography, thin-film deposition, and etching, which are compatible with standard IC batch processing, they exhibit great potential for enabling novel complex systems.

Bulk micromachining is enabled by the process of etching. By judiciously combining highly-directional (anisotropic) etchants, with nondirectional (isotropic) etchants, and the wafer's crystallographical orientation, the etching rates are manipulated to define a wide variety of mechanical structures within the

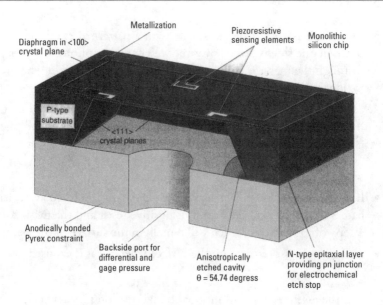

Figure 1.2 A bulk micromachined pressure sensor, shown in cross section, contains a thin silicon diaphragm formed by etching the silicon wafer with alkali-hydroxide. (*From:* [7] © 1994 IEEE. Reprinted with permission.)

confines of the wafer bulk. Furthermore, by creating contours of heavily doped regions, which etch more slowly, and pn junctions, which stop the etching process altogether, it is possible to form deep cavities. Deep cavities are fundamental to the engineering of many devices (e.g., diaphragms for pressure sensors and low loss planar inductors).

Despite its maturity, bulk micromachining has traditionally had some fundamental limitations. For example, the fact that the wafer's crystallographic planes determine the maximum obtainable aspect ratios, restricts attainable device geometry to relatively large sizes as compared with other micromachining technologies [7].

1.3.3 Surface Micromachining

In surface micromachining, thin-film material layers are deposited and patterned on a wafer/substrate. Thin-film material, deposited wherever either an open area or a free-standing mechanical structure is desired, is called *sacrificial* material. The material out of which the free-standing structure is made is called *structural* material. To define a given surface micromachined structure, therefore, a symphony of wet etching, dry etching, and thin-film deposition steps must be composed. Early demonstrations of the potential of surface micromachining were advanced in the 1960s and 1970s by scientists at West-inghouse Electric Corporation, Pittsburgh, Pennsylvania, and at IBM Corpora-tion. At Westinghouse, developments included micromechanical switches and electronic filters that use mechanically resonant thin-film metal structures, as well as advanced light-modulator arrays [8]. At IBM, on the other hand, devel-opments centered on the application of surface micromachining principles to displays, electrostatically actuated mechanical switches, and sensors, in which thin-film oxide structures were integrated with microelectronics [7].

A clearly identifiable point of departure for the development of this tech-nology is the 1967 paper "The Resonant Gate Transistor" [8], which described the use of sacrificial material to release the gate of a transistor. This work dem-onstrated the ability of silicon fabrication techniques to free mechanical systems from a silicon substrate. The next key development on surface micromachining was the use of polysilicon as the structural material, together with silicon dioxide as the sacrificial material, and hydrofluoric acid (HF) to etch silicon dioxide [9].

The next important achievement in the development of surface micromachining technology was its exploitation of structural polysilicon and sacrificial silicon dioxide to fabricate free *moving* mechanical gears, springs, and sliding structures [10, 11]. Since systems applications require that sensors and actuators interface with electronic circuitry, attention turned to the simultane-ous fabrication of micromechanical devices with integrated circuits [12, 13]. Ini-tial devices included polysilicon microbridges [12] and resonant microstructures, which were fabricated together with conventional complemen-tary metal oxide semiconductor (CMOS) and N type MOS processes, respec-tively. Within this context, thin films of polysilicon, grown and deposited silicon dioxide, nitride materials and photoresist, usually provide sensing ele-ments and electrical interconnections, as well as structural, mask, and sacrificial layers. Released mechanical layers have been made with silicon dioxide, alumi-num, polyimide, polycrystalline silicon, tungsten, and single-crystal silicon. Figure 1.3 shows the key steps involved in surface micromachining. Perhaps the most successful surface micromachinng foundry process is that known as multi-user MEM processes (MUMPs).

Alternatives to silicon as a substrate material have been also developed. In the first instance, this came about with the intent of applying surface micromachining techniques to incorporate microwave MEM devices in mono-lithic microwave integrated circuits (MMICs). This entailed exploitation of the semi-insulating low-loss properties of gallium arsenide (GaAs) wafers, and cul-minated in the micromachined microwave actuator (MIMAC) process [14]. In a more recent instance, a surface micromachining process that exploits the low-loss properties alumina substrates and the inexpensive nature of alumina-based microwave integrated circuits (MIC) was also developed [15].

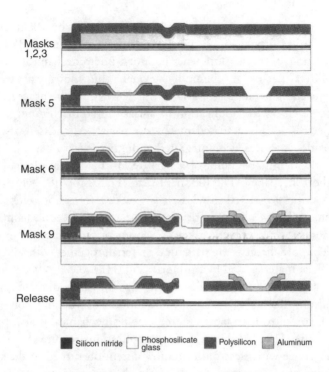

Figure 1-3 In the surface micromachining process, a sacrificial layer is grown or deposited and patterned, and then removed wherever the mechanical structure is to be attached to the substrate. Then the mechanical layer is deposited and patterned. Finally, the sacrificial layer is etched away to release the mechanical structure. (*From:* [7], © 1994 IEEE. Reprinted with permission.)

1.3.3.1   Materials Systems

An examination of the MEMS fabrication literature [16–19] reveals the emergence of a number of self-consistent structural/sacrificial/etchant materials systems. Some examples of these are given in Table 1.1.

Thus, we see that, for example, an aluminum structure may be deposited over photoresist and then set free by released with oxigen plasma; that a polysilicon structure may be deposited over silicon dioxide and set free by release with HF; and that a silicon dioxide structure may be deposited over polysilicon and set free by release with xenon-difluoride.

1.3.3.2   Stiction

The last step in the surface micromachining processing technique is that of dissolving the sacrificial layer to free the structural elements so they can be actuated. This step, however, has been widely reported to be responsible for greatly

Table 1.1
Structural/Sacrificial/Etchant Material Systems

Structural Material	Sacrificial Material	Etchant
Aluminum	Single-crystal silicon	EDP, TMAH, XeF_2
Aluminum	Photoresist	Oxygen plasma
Copper or nickel	Chrome	HF
Polyimide	Aluminum	Al etch (phosphoric, acetic, nitric acid)
Polysilicon	Silicon dioxide	HF
Photoresist	Aluminum	Al etch (phosphoric, acetic, nitric acid)
Silicon dioxide	Polysilicon	XeF_2
Silicon nitride or B-doped polysilicon	Undoped polysilicon	KOH or TMAH

curtailing the yield and reliability of the fabricated MEMS devices due to the phenomenon of stiction [20, 21]. Stiction refers to the sticking of the structural elements either to the substrate or to adjacent elements. It manifests itself in at least two occasions, namely, when the wafer is pulled out of the rinsing solution used to dissolve the sacrificial layer, or when the structural elements find themselves in a humid environment. Many agents have been identified as contributors to stiction. In the first place there is the capillary force of the rinsing liquid, which acts to keep together the released structures and the underlying substrate. As a result of it, both surfaces remain in physical contact even after the postrelease dry. This force is characterized by the liquid surface tension, γ_1 the intersurface distance at which capillary condensation occurs, $d_{0,cap}$, the actual separation between surfaces d, and the contact angle of the liquid and the surfaces, θ [22]. It is given by (1.1),

$$F_{cap} = \frac{2\gamma_l d_{o,cap} \cos \theta}{d^2}$$
(1.1)

In the second place, there are other forms of surface forces, such as electrostatic and van der Waals, which produce permanent adhesion after the system has dried. The electrostatic forces originate from changes on the charge state of adjacent microstructures which elicits a potential difference. The force, characterized by the permitivitty of the medium (air) between the structures, ε_0, the ensuing potential difference, V, and the separation d, is given by (1.2),

$$F_{el}(d) = \frac{\varepsilon_0 V^2}{2d^2}$$
(1.2)

The van der Waals force, in turn, manifests itself when surfaces are in intimate proximity (e.g., separations less than $z_0 \sim 20$ nm), and is characterized by the Hamaker constant, H, and the separation, d, and given by (1.3),

$$F_{vdW}(d) = \frac{H}{6\pi} \frac{z_0}{d^3(d + z_0)} \tag{1.3}$$

Finally, there is solid bridging, which also results in permanent fusion.

A number of antistiction techniques have been advanced to avoid the above issues, and can be classified into three categories. The first category aims at preventing stiction by eliminating the capillary force of the rinsing liquid. This is achieved by either drying the rinsed wafer with supercritical CO_2, or by freezing and then sublimating the rinsing liquid. The second category aims at reducing the structural/substrate contact area, in order to minimize the interface surface energy. This is achieved by introducing reduced contact area structures like dimples and mesas, and by texturizing the substrate (e.g., roughening it). Finally, the third category aims at replacing the liquid rinsing solution by a vapor phase etching (VPE), in which, the sacrificial silicon dioxide is etched with HF vapor instead of the conventional aqueous HF solution.

Stiction considerations are very important in the design of free-standing structures. For example, if the process at hand involves release in an aqueous solution, the thickness and stiffness of cantilever beams must be chosen so that the restoring spring forces can overcome the capillary forces. This selection, in turn, will have ramifications on the nature of the device performance parameters (e.g., the actuation voltage). Figure 1.4 shows a representative [23, 24] structure made via surface micromaching.

Figure 1.4 A representative surface-micromachined component: Slider structure with outer edges guided by self-constraining joints. Stops limit the extent of lateral motion by the slider. (*From:* [11], © 1998 IEEE. Reprinted with permission.)

1.3.3.3 Residual Stress

The state of stress in a differential volume element of a body in equilibrium (seeFigure 1.5), reflects the coexistence of two types of stress, namely, normal stress, σ_i, and shear stress, τ_{ij}, [25]. The normal stress exerts tension or compression on the body, [see Figure 1.5(b)], tending to move its faces parallel to the

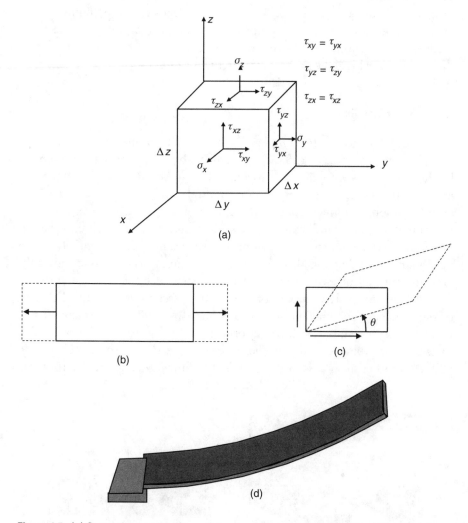

Figure 1.5 (a) State of stress in differential volume element, (b) elongation (dash line) due to tensile normal stress, (c) angular distorsion (dash line) due to shear stress, and (d) sketch of curled-up cantilever beam showing effects of residual stress. Note: Not shown in (b) is the vertical strain, $\varepsilon_{lateral}$ resulting from the longitudinal strain $\varepsilon_{longitudinal}$, via the Poisson effect $\varepsilon_{lateral} = \varepsilon_{longitudinal}$, where ν is Poisson's ratio [25].

faces of the equilibrium differential volume element. The shear stress, on the other hand, tends to distort the equilibrium differential volume element, elongating it along one of its diagonals, [see Figure 1.5(c)]. When an imbalance in the state of normal and shear stresses occurs, the result is likely to be a bent structural element, [see Figure 1.5(d)]. The stress imbalance, in turn, may be caused by differences in thermal expansion coefficients between the material being deposited and the substrate on which it is deposited, or by nonuniformities in its atomic- or grain-level composition, intentional or not, germane to the method of the deposition. of structure. Residual stress is therm used to denote the final stress exhibited by a structure after fabrication [26].

1.3.4 Fusion Bonding

One approach to get greater complexity out of bulk micromachining, is through the separate fabrication of the various elements of a complex system, followed by subsequently assembling them together. Fusion bonding is a technique that enables virtually seamless integration of multiple wafers. This technique relies on the creation of atomic bonds between two wafers either directly or through a thin film of, say, silicon dioxide, as shown in Figure 1.6. The technique had its origin in the development of silicon-on-insulator wafers, and essentially hinges on atomically bonding together two monocrystalline layers to produce much the same mechanical and electrical properties as a single such layer.

As the atomic level nature of the bonding technique requires that it be carried out at high temperatures, approximately 1,100°C for silicon fusion bonding, it is necessary that it be performed before any electrical IC processing to avoid perturbing doping profiles and metallizations. Bryzek, et al. [7], pointed out that to success in this process is a function of surface preparation,

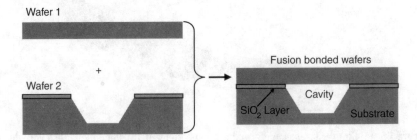

Figure 1.6 Silicon fusion bonding process. A typical approach begins by etching cavities that are later buried beneath the bonded layer. Grinding, etching, and the creation of the electrical devices follow this. Bulk micromachining and finally a release etch descending to the buried cavities complete the process. (*From:* [7], © 1994 IEEE. Reprinted with permission.)

and that, while still an emerging technology, it has great potential as evidenced by the appearance of various commercial devices, namely, pressure sensors, accelerometers, and resonant structures.

While not necessarily amenable to batch processing, fusion bonded micro-structures do compare favorably with respect to those obtained by surface and bulk techniques. For example, there is a demonstrable increase in the maximum achievable release layer undercut, which is an order of magnitude thicker than that possible with surface micromachining, and it also enables the design and fabrication of structures which would otherwise be impractical using bulk techniques alone.

1.3.4.1 SOI-MEMS Process

The silicon-on-insulator (SOI) process is typically employed in applications where the stress-free and high (mechanical) quality factor of single-crystal silicon are key to device performance. Such is the case in MEM resonators [27]. An example of the process flow utilized by Ayazi [27], is found in Figure 1.7.

After opening windows on the top oxide layer, [see Figure 1.7(a)], the single-crystal silicon structural layer is etched, with the depth delimited by the bonding oxide, which works as an etch stop. Subsequently, the bonding oxide is etched, resulting in the creation of a gap or cavity underneath the beam.

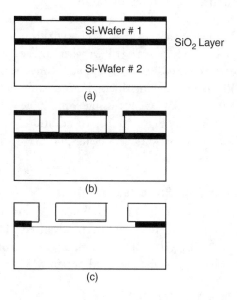

Figure 1.7 Steps of SOI-MEMS process. (a) Windows are opened on the top oxide layer. (b) The single-crystal-silicon structural layer (beam) is etched. (c) The bonding oxide is etched to create a cavity underneath the beam. (*After:* [27].)

1.3.5   LIGA and Sacrificial LIGA

While bulk micromachining sculpts three dimensional structures in the bulk of a substrate via the ingenious application of isotropic and anisotropic etchants, highly doped layers, and pn junctions, and surface micromachining does it via the sequential deposition, patterning, and release of thin film layers, LIGA does it by first creating a mold of the desired structure itself on the substrate surface [28, 29]. The creation of the desired three dimensional structure is formed in three steps. First, a metal substrate is coated with a thick layer of photoresist, on which the mask (the mold footprint) is patterned and developed. The thickness of the photoresist defines the aspect ration of the resulting mold. To penetrate thick layers of photoresist, and thus achieve large aspects ratios, it is necessary to provide exposure to high energy synchrotron X-ray radiation. Secondly, the pattern thus created on the photoresist is used to pattern plate a mold insert (seeded by the metal substrate). And thirdly, the mold insert structure, which can be conductive or magnetic, is used to effect reactive injection molding, thermoplastic injection molding, or embossing of the desired structure. Polymers, such as polyimide, polymethyl methacrylate (PMMA) are typically engraved by these molds. Since the process creates the mold patterns by deep X-ray lithography (i.e., capable of very small resolution), the structures achievable have large vertical aspect ratios, lateral dimensions of a few micrometers, and vertical dimensions up to 1,000 μm, as shown in Figure 1.8. Figure 1.9 shows some representative components fabricated by LIGA.

Despite its unique pattern-forming properties, the utilization of LIGA is limited by the need of having access to an X-ray synchrotron facility. A compromise, which combines some features of LIGA with surface micromachining and, thus, obviates the need for exposure to X-rays has been devised [7], namely, sacrificial (S) LIGA [30]. It replaces the thick photoresist by polyimide as the electroplating mold, so that it permits processing compatible with conventional IC batch processes, as shown in Figure 1.10. In addition, by effecting multiple coatings, it is possible to increase the resulting photoresist mold thickness and aspect ratio [30].

1.3.6   Hybrid Micromachining Processes

To further improve the versatility of micromachined structures, a number of creative micromachining processes have been advanced. These include, deep reactive ion etching (DRIE) [31, 32], the single crystal silicon reactive etch and metal (SCREAM) process [33], and stereolithography [34–36].

1.3.6.1 Bulk Micromachining by DRIE

In DRIE, (see Figure 1.11), silicon is etched anisotropically, by repeating a sequence of separate plasma etching and polymirization steps [31, 32]. In plasma etching, positive ions resulting from the breakdown discharge of a gas

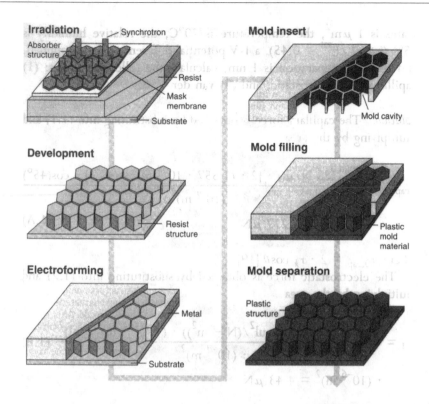

Figure 1.8 LIGA process flow. Metal is selectively plated onto thick photoresist overlaying a conductive substrate and patterned by exposure to X-rays. In this way, metal structures up to 1 mm high and only a few microns wide are created. (*From:* [7], © 1994 IEEE. Reprinted with permission.)

above the silicon wafer, bombard the silicon surface as they fall vertically towards the negatively charged wafer. To avoid lateral etching and, thus achieve vertical selectivity, the sidewalls are protected by a polymer (PR). Upon doing so, etching is effected at the bottom of the trench. Each etching step is stopped after the maximum tolerated lateral etch is produced. This lateral etch can be of the order of tenths of microns. By repeating the passivation/etch sequence, trenches with overall depths of up to several hundred microns have been demonstrated. The process proceeds at room temperature, can produce selectivities of 200:1 in standard PR masks, 300:1 in hard masks such as SiO_2 and Si_3N_4, and exhibits etching rates of $6\,\mu$m/sec [31].

1.3.6.2 Bulk Micromachining by SCREAM

The single crystal silicon reactive etch and metal (SCREAM I) process effects bulk micromachining using plasma and reactive ion etching (RIE) [33], as

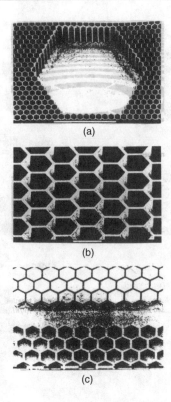

Figure 1.9 Representative components fabricated by LIGA: (a) honeycomb structures, and (b) self-supporting Fresnel zone plate. (*From:* [28] © 1987 IEEE. Reprinted with permission.)

shown in Figure 1.12. The process is self-aligned, employs one mask to define structural elements and metal contacts, and its temperature does not exceed 300°C, while using standard tools. The low temperature requirement makes it amenable for integration of MEMS devices with very large scale integration (VLSI) technology [33].

1.3.6.3   Stereolithography

In this technique, (see Figure 1.13), a three-dimensional body is constructed on a layer-by-layer basis. The main elements of the system to accomplish this are: two stages with two-dimensional (XY) and one-dimensional (Z) motion, or one stage with three-dimensional motion, a tank containing photocurable resin, and an ultraviolet laser beam source.

The spot of the UV beam strikes the resin perpendicularly, and it is chosen to have such an intensity and duration that it penetrates and cures a certain

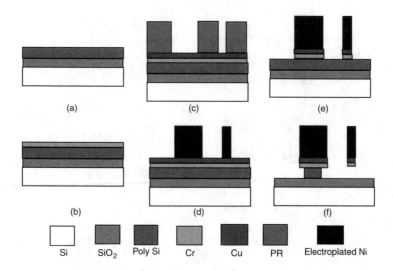

Figure 1.10 Process flow of high-aspect-ratio electroplating process. (a) LPCVD Si02~2 μm, LPCVD Si02~2 μm; (b) seed layer deposition, Cu~1,000Å, Cr~50Å; (c) PR mold lithography; (d) Plate Ni, remove PR mold; (e) seed layer etching; (f) polysilicon etching by XeF$_2$. (*After:* [30].)

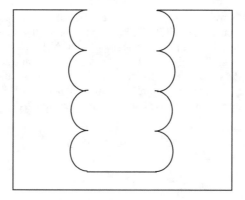

Figure 1.11 Sketch of trench etched by DRIE. Scalloping is shown (exaggerated) on the sidewalls reflecting the passivation-etching sequence.

depth Δz of the resin. As the stage on which the resin-containing tank rests is moved in XY (or as the computer-control laser beam traces the shape), a particular cross section of the body, as approximated by a flat slice of thickness Δz, is drawn and cured. Next, lowering the stage a distance Δz, another amount of liquid resin of depth Δz is ready to be exposed and cured/hardened. In state of the art systems, the vertical layer resolution Δz may be as small as a few microns [34–36]. The process is thus repeated until the whole body is constructed.

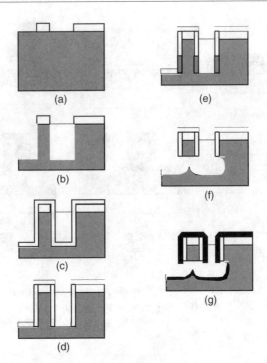

Figure 1.12 SCREAM I process flow: (a) Deposition and patterning of PECVD masking oxide, (b) RIE of silicon with BCl_3/CL_2 RIE, typically 4 to 20 μm deep, (c) deposition of oxide sidewall via PECVD, typically 0.3 μm thick, (d) vertical etch of bottom oxide with CF_4/O_2 RIE, (e) etch of silicon 3 to 5 μm beyond end of sidewall with Cl_2 RIE, (f) isotropic RIE release of structures with SF_6 RIE, and (g) sputtering deposition of aluminum metal. The device shown is a beam, free to move left-right, and its corresponding parallel-plate capacitor. (*After:* [33].)

1.4 Exercises

Exercise 1.1 Order of Magnitude of Stiction Forces

Stiction forces develop, typically, either as a result of the release process or because of the proximity of flat surfaces in a humid environment. Whatever the case, the situation may be illustrated as in Figure 1.14.

Assume there is the potential for water to condensed between the two perfectly flat and parallel plates of silicon, with a contact angle, $\theta = 45°$, and a surface tension [37] $\gamma_l = 72.88$ dynes/cm $= 0.07288$ N/m. Then if the area of the plates is 1 μm^2, the temperature is 20°C, the relative humidity is 45% (i.e., $P/P^{sat} = 0.45$), a 1-V potential difference is being applied, and the plate separation is 1 nm, calculate the following forces: (a) capillary, (b) electrostatic, and (c) van der Waals.

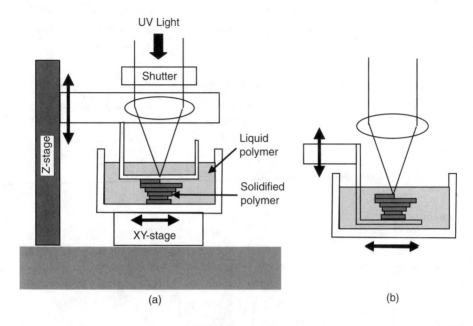

Figure 1.13 (a) Conventional "1H" process for stereolithography, and (b) super "1H" or free-form stereolithography. (*After:* [34, 35].)

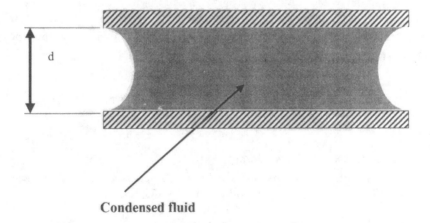

Figure 1.14 Illustration of two flat surfaces being pulled together by capillary forces of condensed liquid $r_k = \gamma_l v / RT \log(P/P^{sat})$, $\gamma_l v / RT \approx 0.54\, nm$ for water at 20°C. (*After:* [17].)

Solution:

(a) Capillary force

The capillary force is obtained substituting into (1.1) and multiplying by the area:

$$F_{cap} = \frac{2\left(0.07288\,\frac{N}{m}\right)\cdot\left[2\cdot\left(1.557\cdot10^{-9}\,m\right)\cos(45°)\right]\cdot\cos(45°)}{\left(10^{-9}\,m\right)^2}$$

$$\cdot\left(10^{-6}\,m\right)^2 = 227\ \mu N$$

where $d_{0,cap} = 2\ r_k\cos\theta$[22].

(b) Electrostatic force

The electrostatic force is obtained substituting into (1.2) and multiplying by the area:

$$F_{el} = \frac{\left(8.85418\cdot10^{-12}\,\frac{coul^2}{N-m^2}\right)\cdot\left(1\,\frac{N-m}{coul}\right)^2}{2\cdot\left(10^{-9}\,m\right)^2}\cdot\left(10^{-6}\,m\right)^2 = 4.43\ \mu N$$

(c) van der Waals force

Using the Hamaker constant for silicon [22], $H = 1.6$ eV $= 2.5744\cdot10^{-19}$ N$-$m, the van der Waals force is obtained substituting into (1.3) and multipliying by the area:

$$F_{vdW} = \frac{\left(2.5744\cdot10^{-19}\,N-m\right)\cdot\left(20\cdot10^{-9}\,m\right)}{6\pi\left(10^{-9}\,m\right)^3\cdot\left(10^{-9}\,n+20\cdot10^{-9}\,m\right)}\cdot\left(10^{-6}\,m\right)^2 = 13\ \mu N$$

Exercise 1.2 Dominion of Stiction Forces

Assuming the same conditions as above, for a plate separation of 100 nm, calculate (a) capillary, (b) electrostatic, and (c) van der Waals forces.

Solution:

Changing the parameter d in the above calculations to $d = 100$ nm, one obtains:

(a) $F_{cap} = 2,270\ \mu\mu N$
(b) $F_{el} = 443\ \mu\mu N$
(c) $F_{vdW} = 2.3\ \mu\mu N$

Notice that the nature of the dominant stiction force varies as a function of the interplate distance d.

1.5 Summary

In this chapter we have discussed the origins of and motivation behind the development of MEMS, and reviewed the techniques germane to the fabrication of MEMS. In particular, we have described the evolution of micromachining techniques, which were originally motivated by the need to simultaneously fabricate sensors and actuators together with their interface circuitry. We started by introducing the conventional two-dimensional pattern definition techniques of conventional IC technology, and went on to describe techniques which enable the creation of three-dimensional structures on a planar substrate, namely, bulk micromachining, surface micromachining, fusion bonding, LIGA and SLIGA, and hybrid techniques. This included the illustration of microstructures typical of each fabrication technique, as well as of the techniques' limitations in terms of yield and reliability.

In the next chapter we will deal with the fundamental physics that govern MEM devices operating via electrostatic actuation, the actuation mechanism most often employed in the realization of microwave MEM devices.

References

[1] Feynman, R. P., "There's Plenty of Room at the Bottom," presented at the American Physical Society Meeting in Pasadena, CA, December 26, 1959; reprinted with permission of Van Nostrand Reinhold in *J. Microelectromechanical Systems*, Vol. 2, 1992, pp. 60–66.

[2] Feynman, R. P., "Infinitesimal Machinery," presented at the Jet Propulsion Laboratory on February 23, 1983; reprinted in *J. Microelectromechanical Systems*, Vol. 1, 1993, pp. 4–14.

[3] Senturia, S. D., "Feynman Revisited," *IEEE Micro Electro Mechanical Systems Workshop*, Oiso, Japan, January 25–28, 1994, pp. 309–312.

[4] Hirano, M., H. Kuwano, and J. W. Weijtmans, "Superlibricity Mechanism for Microelectromechanical Systems," *IEEE Microelectromechanical Systems Workshop*, Nagoya, Japan, 1997, pp. 436–441.

[5] Bacher, W., W. Menz, and J. Mohr, "The LIGA Technique and Its Potential for Microsystems—A Survey," *IEEE Trans. Ind. Electronics*, Vol. 42, No. 5, 1995, pp. 431–441.

[6] Jaeger, R. C., *Introduction to Microelectronics Fabrication*, Volume V, Modular Series on Solid State Devices, G. W. Neudeck and R. F. Pierret, (eds.), Boston, MA: Addison-Wesley, 1988.

[7] J. Bryzek, K. Petersen, and W. McCulley, "Micromachines on the March," *IEEE Spectrum*, 1994, pp. 20–31.

[8] Nathanson, H. C., et al., "The Resonant Gate Transistor," *IEEE Trans. Electron Dev.*, Vol. 14, No. 3, 1967, pp. 117–133.

[9]   Howe, R. T., and R. S. Muller, "Polycrystalline Silicon Micrimechanical Beams," *J. Electrochemical Soc.*, Vol. 130, 1983, pp. 1420–1423.

[10] Mehregany, M., K. J. Gabriel, and W. S. N. Trimmer, "Integrated Fabrication of Polysilicon Mechanisms," *IEEE Trans. Electron Dev.*, Vol. 35, No. 6, 1988, pp. 719–723.

[11] Fan, L.-S., Y.-C. Tai, and R. S. Muller, "Integrated Movable Micromechanical Structures for Sensors and Actuators," *IEEE Trans. Electron Dev.*, Vol. 35, No. 6, 1988, pp. 724–730.

[12] Parameswaran, M., H. P. Baltes, and A. M. Robinson, "Polysilicon Microbridge Fabrication Using Standard CMOS Technology," *Tech. Digest, IEEE Solid-State Sensor and Actuator Workshop*, Hilton Head Island, South Carolina, June 6–9, 1988, pp. 148–150.

[13] Putty, M. W., et al., "Process Integration for Active Polysilicon Resonant Microstructures," *Sensors & Actuators*, Vol. 20, 1989, pp. 143–151.

[14] Larson, L. E., R. H. Hackett, and R. F. Lohr, "Microactuators for GaAs-Based Microwave Integrated Circuits," *IEEE Conference on Solid State Sensors and Actuators*, Hilton Head Island, South Carolina, 1991, pp. 743–746.

[15] De Los Santos, H. J., et al., Microelectromechanical Device: US patent # US6040611, Issued: March 21, 2000, Assignee: Hughes Electronics.

[16] Williams, K. R., and R. S. Muller, "Etch Rates for Micromachining Processing," *J. Microelectromechanical Syst.*, Vol. 5, No. 4, 1996, pp. 256–259.

[17] Schmidt, M. A., et al., "Design and Calibration of a Microfabricated Floating-Element Shear-Stress Sensor," *IEEE Trans. Electron Dev.*, Vol. 35, No. 5, 1988, pp. 750–757.

[18] Tabata, O., et al., "Surface Micromachining Using Polysilicon Sacrificial Layer," *The Second Int. Symp. on Micromachine and Human Science*, Tokyo, Japan, 1991.

[19] Frazier, A. B., and M. G. Allen, "High Aspect Ratio Electroplated Microstructures Using a Photosensitive Polyimide Process," *Proc. IEEE MEMS 92*, Travemunde, Germany, February 1992, pp. 87–92.

[20] Kim, C. J., et al., "Polysilicon Surface-Modification Technique to Reduce Sticking of Microstructures," *Sensors and Actuators*, Vol. A 52, 1996, pp. 145–150.

[21] Lee, Y. Iet al., "Dry Release for Surface Micromachining with HF Vapor-Phase Etching," *J. Microelectromechanical Syst.*, Vol. 6, No. 3, 1997, pp. 226–233.

[22] Maboudian, R., and R. T. Howe, "Critical Review: Adhesion in Surface Micromechanical Structures," *J. Vac. Sci. Technol.*, Vol. B15, No. 1, 1997, pp. 1–20.

[23] Tang, W. C., M. G. Lim, and R. T. Howe, "Electrostatically Balanced Comb Drive for Controlled Levitation, " *Tech. Digest Solid-State Sensor and Actuator Workshop*, Hilton Head Island, South Carolina, 1990, pp. 23–27.

[24] Macdonald, N. C., et al.,"Selective Chemical Vapor Deposition of Tungsten for Microelectromechanical Structures" *Sensors and Actuators*, Vol. 20, 1989, pp. 123–133

[25] Higdon, A., et al., *Mechanics of Materials*, New York: John Wiley & Sons, Inc, 1976.

[26] Senturia, S. D., *Microsystem Design*, Boston, MA: Kluwer Academic Publishers, 2001.

[27]   Piazza, G., R. Abdolvand, and F. Ayazi, "Voltage-Tunable Piezoelectrically-Transduced Single-Crystal Silicon Resonators on SOI Substrate," *2003 IEEE MEMS Comference*, Kyoto, Japan, January 19–23, 2003, pp. 149–152.

[28]   Ehrfeld, W., et al., "Fabrication of Microstructures Using the LIGA Process," *Proc. IEEE Micro Robots and Teleopeartos Workshop*, Hyannis, MA, November 1987.

[29]   Frazier, A. B., R. O. Warrington, and C. Friedrich, "The Miniaturization Technologies: Past, Present, and Future," *IEEE Trans. Ind. Electronics*, Vol. 42, No. 5, 1995, pp. 423–431.

[30]   Li, X., et al., "High-Aspect-Ratio Electroplated Structures with 2 m Beamwidth," *Proc. MEMS (MEMS-Vol. 1), ASME Int. Mechanical Engineering Congress and Exposition*, Nashville, TN, November 1999, pp. 25–30.

[31]   Laermer, F., and A. Schlip, "Method of Anisotropically Etching Silicon," Patent # 5501893, Issued: March 26, 1996, Assignee: Robert Boch GmbH, Stuttgart, Germany.

[32]   Laermer, F., et al., "Bosch Deep Silicon Etching: Improving Uniformity and Etch Rate for Advanced MEMS Applications," Proc. IEEE MEMS Workshop, Orlando, FL, Jan 17–21, 1999.

[33]   Shaw, K. A, Z. L. Zhang, and N. C. MacDonald, "SCREAM I: A Single Mask Single-Crystal Silicon Process for Microelectromechanical Structures," *Sensors and Actuators*, A 40, 1994, pp. 210–213.

[34]   K. Ikuta, and K. Hirowatar, "Real Three Dimensional Micro Fabrication Using Stereo Lithography and Metal Molding," *IEEE Proc. MEMS Conf.*, Fort Lauderdale, FL, 1993, pp. 42–47.

[35]   K. Ikuta, et al., "New Micro Stereo Lithography for Freely Movable 3D Micro Structure," *IEEE Proc MEMS Conf.*, Heidelberg, Germany, January 1998, pp. 290–295.

[36]   A. S. Holmes, "Laser Fabrication and Assembly Processes for MEMS," *Proc. LASE, High-Power Lasers and Applications*, San Jose, CA, January 20–26, 2001.

[37]   Vargaftik, N. B., *Handbook of Physical Properties of Liquids and Gases: Pure Substances and Mixtures*, Second Edition, Washington, D.C.: Hemisphere Publishing Corporation, 1975.

2

Fundamental MEMS Device Physics

2.1 Actuation

Actuation refers to the act of effecting or transmitting mechanical motion, forces, and work by a device or system on its surroundings, in response to the application of a bias voltage or current. A wide variety of actuation mechanisms have been researched in the MEMS field [1, 2]. These include electrostatic, piezoelectric, electromagnetic, shape memory alloy [(SMA), that is, materials which, upon experiencing deformation at a lower temperature, can return to their original undeformed shape when heated], and thermoelectromechanical [2]. In view of the many mechanisms available for exploitation in the engineering of actuation-based devices, it is natural to ask whether there are some to be preferred over others when it comes to their application in microelectromechanical microwave systems. As the MEMS field is relatively new, this question has perhaps only been answered in the affirmative in what pertains to one of the above mechanisms, namely, electrostatic. Indeed, it has been found that at microscopic scale sizes it is easier to produce electric fields [3, 4]. For example, electric fields exceeding 3×10^6 V/m, the coronal discharge of air, can be easily generated by applying low voltages across the micron-sized gaps encountered in micromachined ICs. Moreover, even nature seems to be teaching us that electrostatic fields are best suited for producing actuation in small devices; muscular motion is the response to electrostatic forces [3].

The fact that surface micromachining, the most common technology utilized to produce electrostatically based actuators, is compatible with integrated circuit fabrication processes has been a motivating force behind the strong interest in developing complex microsystems whose micromechanical component relies on electrostatic actuation. Other innate advantages of electrostatic actuators include the inherent simplicity of their design, their fast response, their ability to achieve rotary motion, and their low power consumption [1]. When examined in light of their potential for compatibility with integrated electronic functions, all the other actuation mechanisms mentioned above fall short for two primary reasons. The first reason is that one finds that their fabrication requires specialized materials and processing not found in standard IC manufacturing. Under this category one has, for instance, electromagnetic actuators, which require multiturn magnetic windings and their accompanying metallic cores, and piezoelectric actuators, which require the deposition of exotic materials, like lead-zirconate-titanate (PZT). Secondly, one finds that the level of performance is substandard. This category is exemplified by SMA and thermoelectromagnetic actuators, which are power-inefficient and display slow reaction rates. Against this backdrop, then, it is clear that, at this juncture in the development of the MEMS field, electrostatically based actuation devices are bound to remain the main candidate for microelectromechanical microwave systems. Despite this fact, and given recent developments which might make their utilization more prominent in the next few years, the fundamentals of the heat-driven, piezoelectric, and magnetic actuators have been also addressed in this chapter.

2.1.1   Electrostatic Actuation

2.1.1.1   Parallel-Plate Capacitor

Consider a parallel-plate capacitor, in which the plates are rigid and constrained from moving, as shown in Figure 2.1. Assuming the area of the plates is much

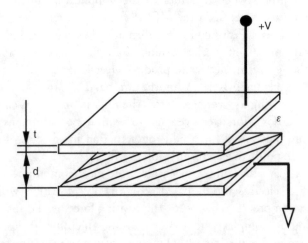

Figure 2.1 Parallel-plate capacitor as an actuator. V is the applied voltage, d is the plate separation, t its thickness, and ε is the permittivity of the volume between the plates. (*After*: [2].)

greater than the separation between them (i.e., ignoring the fringing fields), its capacitance is given by (2.1) [2]:

$$C = \frac{\varepsilon A}{d} \tag{2.1}$$

where ε denotes the dielectric permittivity of the medium between the plates, d represents the distance separating the plates, and A represents the area of the plates. Corresponding to a voltage V applied to the capacitor there exists an electrostatic potential energy, given by (2.2), stored in the volume between the plates.

$$U = \frac{1}{2}CV^2 \tag{2.2}$$

This potential energy represents the energy required to prevent the oppositely charged parallel plates from collapsing into each other as a result of the coulomb force of attraction, (2.3),

$$F = \frac{1}{4\pi\varepsilon}\frac{q_T q_B}{d^2} \tag{2.3}$$

where q_B and q_T are the equal but opposite charges in the bottom and top plates respectively. Alternatively, this force may also be expressed as the negative of the gradient in the potential energy between the parallel plates, (2.4).

$$F = -\nabla U \tag{2.4}$$

Substituting (2.1) into (2.2), we obtain (2.5).

$$U = \frac{\varepsilon A V^2}{2d} \tag{2.5}$$

Now substitution of (2.5) into (2.4) yields

$$F = \frac{\varepsilon A V^2}{2d^2} \tag{2.6}$$

This equation quantifies the force that must be applied on the top plate in order to prevent it from collapsing on the bottom plate if the top plate were freed. It expresses that this force increases linearly with area, quadratically with voltage, and decreases quadratically with the separation between plates.

Suppose now that, while the applied voltage is maintained at V, the top plate is suddenly unconstrained. Then, as the top plate is now free to move, the coulomb force of attraction will make it approach the bottom plate (i.e., the gap d will decrease). If the gap d decreases, the capacitance, in turn, increases by (2.1). But if the capacitance increases, then the stored energy also increases by (2.2). This rate of increase in stored potential energy, which is caused by the rate of decrease in gap spacing gives, by (2.4), the instantaneous force of attraction between the plates, which further drives the gap closure. The feedback process just described would ultimately culminate in the condition of zero gap (i.e., collapse of the plates). We see, therefore, that the application of a voltage to a parallel-plate capacitor may result in the motion or actuation of its plates.

If we lift the condition that the top plate be rigid, and assume that it is anchored at only one of its four sides, the top plate may be considered as a beam, namely, a cantilever beam, as shown in Figure 2.2. Then, in response to an applied voltage V, the top plate will deflect with zero deflection at the anchoring juncture, to a maximum deflection at the tip of the beam.

As easily deduced from Figure 2.2, by controlling the state of deflection of the beam, switching may be effected. The fundamental physics characterizing electrostatic actuation is embodied in the relationship between the deflection and the applied voltage causing it.

Peterson [5] obtained an approximate deflection-voltage relationship by modeling the cantilever beam as a parallel-plate capacitor whose top plate experiences a distributed force, as shown in Figure 2.3. As this force varies along the length of the beam, the interelectrode gap d becomes a function of the length. Taking this gap variation into account in (2.6), the electrostatic force exerted on the beam at a point x due to the electrostatic potential is given by (2.7).

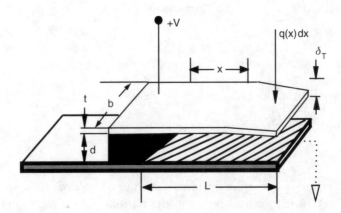

Figure 2.2 Cantilever beam structure indicating geometrical parameters used in analysis. (*After:* [5].)

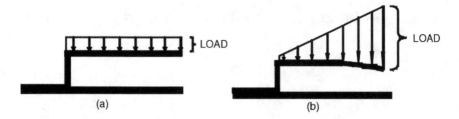

Figure 2.3 Electrostatic load on cantilever beam (a) Initial uniform load, and (b) nonuniform load after displacement.

$$F = \frac{\varepsilon A V^2}{2(d - \delta(x))^2} \tag{2.7}$$

As one would expect, the beam deflection $\delta(x)$ resulting from a given electrostatic force is a function of the structural and material properties of the beam, as given by its Young's modulus, E, and its moment of inertia, I. In particular, a concentrated load at a position x on a cantilever beam results in a deflection at the beam tip given by (2.8),

$$\delta_T = \left[\frac{x^2}{6EI} \right] (3L - x) b q(x) \, dx \tag{2.8}$$

where b is the beam width and

$$q(x) = \frac{\varepsilon}{2} \left(\frac{V}{d - \delta(x)} \right)^2 \tag{2.9}$$

The forces are distributed along the length of the beam, and the deflection at the tip is found by integrating (2.8) from $x = 0$ to $x = L$, (2.10).

$$\delta_T = b \int_0^L \frac{3L - x}{6EI} x^2 q(x) \, dx \tag{2.10}$$

Petersen assumed that the beam deflection at any point x along the beam could be approximated by a square-law dependence, namely,

$$\delta(x) = \left(\frac{x}{L} \right)^2 \delta_T \tag{2.11}$$

and the integral in (2.10) was solved to give a normalized load, defined in (2.12),

$$l = \frac{\varepsilon b L^4 V^2}{2 E I d^3} \tag{2.12}$$

required to give a normalized deflection at the tip of the cantilever beam given by (2.13).

$$\Delta = \frac{\delta_T}{d} \tag{2.13}$$

Expressed in terms of Δ, the normalized load obtained upon evaluating the integral in (2.10) is given by (2.14).

$$l = 4\Delta^2 \left[\left(\frac{2}{3(1-\Delta)} \right) - \frac{\tanh^{-1} \sqrt{\Delta}}{\sqrt{\Delta}} - \frac{\ln(1-\Delta)}{3\Delta} \right]^{-1} \tag{2.14}$$

A plot of (2.14), as shown in Figure 2.4, reveals that there exists a threshold bias at which the beam snaps (i.e., is no longer controlled by the applied bias), and comes crashing down on the substrate. This bias point is reached after

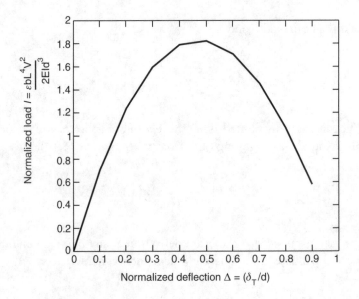

Figure 2.4 Normalized deflection/normalized load characteristic for cantilever beam. (*From:* [5], © 1978 IEEE. Reprinted by permission).

the beam deflects about one-third of the original beam-to-electrode distance, as explained by Petersen. He noted that, as the beam bends downward, the electrostatic forces become increasingly concentrated at the tip, so that at a particular voltage, the concentrated load causes the beam position to become unstable and it undergoes a spontaneous deflection, the remaining distance.

An approximate expression for this threshold voltage, valid for a homogeneous beam, is given by (2.15).

$$V_{th} = \sqrt{\frac{18EId^3}{5\varepsilon L^4 b}} \qquad (2.15)$$

The occurrence of pull-in behavior is also revealed by an examination of the equation of equilibrium between spring and electrostatic forces whose solution, as a function of applied voltage, gives the displacement of the movable plate of a parallel-plate capacitor, $x(V)$, (see Figure 2.5). This equation is given by:

$$\left(kx - \frac{\varepsilon_0 A V^2}{2(d_0 - x)^2} \right) = 0 \qquad (2.16)$$

and it implies that, beginning from $V = 0^+$, x will adjust itself so that (2.16) is *eventually* satisfied. This adjustment in x will only occur, however, as long as the magnitude of V is such that the *initial* value of (2.16) is greater than zero. The implication, of course, is that it takes a small but finite time for (2.16) to be satisfied once V changes. In that case, there will be positive real solutions [i.e., roots, $x(V) > 0$]. What is the maximum voltage at which it will no longer be possible to find an x that satisfies (2.16)? This is the voltage at which the difference between spring and electrostatic forces, embodied by (2.16), is a minimum. Its value may be found by calculating the derivative of (2.16) at equating it to zero,

$$\frac{dV}{dx} = \frac{d}{dx}\left[\sqrt{\frac{2kx(d_0 - x)^2}{\varepsilon_0 A}} \right] = 0 \rightarrow x = \frac{d_0}{3}. \qquad (2.17)$$

Then, substituting $x = d_0/3$ into (2.16) and solving for V, one obtains,

$$V_{Pi} = \sqrt{\frac{8kd_0^3}{27\varepsilon_0 A}} \qquad (2.18)$$

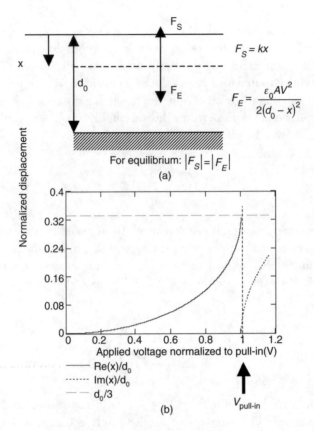

Figure 2.5 (a) Force diagram for parallel-plate capacitor in equilibrium, and (b) normalized displacement versus applied voltage normalized to pull-in voltage.

This is the pull-in voltage. A plot of the normalized displacement versus the applied voltage normalized to the pull-in voltage [see Figure 2.5(b)], shows that at pull-in, the displacement becomes imaginary and the slope of the applied voltage no longer controls the displacement.

The operation of the cantilever beam, which includes it being driven into the instability regime, is called *hysteretic*. Once the beam is fully deflected, subsequent reduction of the applied voltage will have no effect on its state of deflection (neglecting any stiction effects) until it becomes lower than the threshold. Zavracky, Majumder, and McGruer [6] have derived concise expressions for the beam closing (threshold) and opening voltages in terms of the effective beam spring constant, $K = bt^3 E/4l^3$, the beam area A, and the initial and effective closed beam-to-electrode distances, d and d_e, respectively. Accordingly, the closing and opening voltages are given by (2.19) and (2.20), respectively.

$$V_{th-close} = \frac{2}{3} d \sqrt{\frac{2Kd}{3\varepsilon_0 A}} \tag{2.19}$$

$$V_{th-open} = (d - d_c) \sqrt{\frac{2Kd_c}{3\varepsilon_0 A}} \tag{2.20}$$

In some applications it is desirable to avoid losing control of the beam deflection. This requires that the beam bias voltage be kept from exceeding the threshold for spontaneous deflection.

While the parallel-plate capacitor is a simple actuation device, it possesses one important drawback; that is, the force of actuation drops off too fast with increasing gap (i.e., $F \propto \frac{1}{d^2}$). This relationship is undesirable because the usable range of displacement is limited. The difficulty, however, is surmounted by the interdigitated "comb-drive" capacitor, the subject of the next section.

2.1.1.2 Interdigitated "Comb-Drive" Capacitor

A key drawback of the parallel-plate capacitor as an actuator is the rapidity with which the force it exerts drops off with increasing gap. This difficulty is overcome by the interdigitated comb-drive capacitor [7], as shown in Figure 2.6.

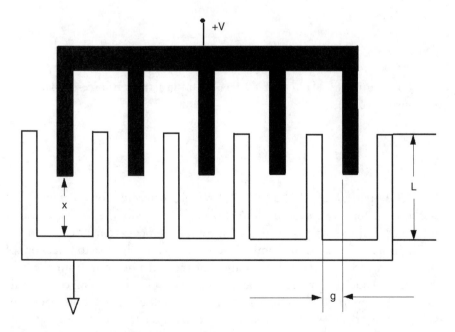

Figure 2.6 Interdigitated comb drive actuator. The thickness into the plane is *t*. (*After:* [2].)

To obtain the voltage-displacement relationship for the electrostatic comb-drive, we assume that both the upper and lower elements are constrained from moving, and that the upper element is held at a fixed voltage V [2]. Then we apply (2.1) to (2.5), with the capacitance for a single tooth face across the gap given by (2.21),

$$C_{single} = \frac{\varepsilon A}{g} \tag{2.21}$$

where the area is given by (2.22).

$$A = t(L - x) \tag{2.22}$$

The capacitance then is given by (2.23).

$$C_{single} = \frac{\varepsilon t(L - x)}{g} \tag{2.23}$$

Since each tooth has two sides, it follows that each tooth has two capacitors. For an n teeth upper actuator, we have $2n$ capacitors and hence the total capacitance is,

$$C_{single} = 2n \frac{\varepsilon t(L - x)}{g} \tag{2.24}$$

Substituting (2.24) into (2.4), we obtain the actuation force-displacement relationship, (2.25).

$$F = n\varepsilon \frac{t}{g} V^2 \tag{2.25}$$

A comparison of (2.6) and (2.25) reveals that while for a parallel-plate capacitor the force varies as $1/x^2$, for the comb-drive device the force is constant independent of x as long as the degree of comb finger engagement is reasonable [2, 7]. Fringing fields for the comb-drive actuators also give rise to forces out of the plane, which can result in levitation of the actuator away from the substrate [2]. In addition, there will be a lateral instability depending on how the actuator is supported. If the lateral stiffness is insufficient, the upper actuator will be attracted sideways and the upper actuator teeth will stick to the fixed teeth [2].

2.1.2 Piezoelectric Actuation

While the electrostatic mechanism of actuation relies on the attraction or repulsion between charges distributed on the surface of structures to effect the mechanical motion of these structures, piezoelectric actuation relies on the deformation of structures caused by the motion of *internal* charges as a result of an applied electric field. Conversely, an applied stress on a piezoelectric structure elicits an electric field in it as a result of the forced motion of the internal charges. Due to the anisotropic properties of piezoelectric crystals, there is coupling between electric fields and strains in different directions. For example, in a crystal where a z-directed electric field may produce a strain in the x-direction, this behavior is captured by the following set of constitutive equations [8]:

$$S_x = s_{xx}^E T_x + d_{zx} E_z$$

$$D_z = d_{zx} T_x + \varepsilon_{zz}^T E_z$$

(2.26)

where S_x is the strain in the x-direction, s_{xx}^E is the elastic compliance in the x-direction in the absence of an electric field due to an x-directed stress, T_x is an x-directed stress, d_{zx} is the piezoelectric coefficient relating a z-directed electric field E_z field to an x-directed strain, ε_{zz}^T is the z-directed permittivity at constant stress due to a z-directed electric field, and D_z is the electric displacement.

2.1.2.1 Cantilever Probes

The fundamental piezoelectric MEMS device is the cantilever beam, as shown in Figure 2.7. In this case the beam consists of a composite-layer structure of a

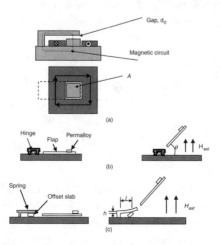

Figure 2.7 Fundamental piezoelectric MEMS device. (*After:* [9].)

piezoelectric material sandwiched between two electrodes (i.e., a capacitor). A voltage across the capacitor sets an electric field in the z-direction, which causes a strain/elongation of the piezoelectric layer in the x-direction. Since the beam is not piezoelectric, its size does not change and the result is bending (i.e., its tip displaces in the z-direction). The magnitude of the displacement is a function of the lateral width of the beam, with a value of several microns per 100 μm of beam width being typical, together with a change in beam thickness of 0.1 nm per volt [9]. Typical piezoelectric materials employed in MEMS devices include zinc oxide (ZnO), aluminum nitride (AlN), lead zirconate titanate (PbZrxTi$_{1-}$xO$_3$- or more commonly, PZT), and polyvinyledene fluoride (PVDF).

A detailed analysis of piezoelectric multimorph actuation motion of a cantilever beam was given by Weinberg [10]. Results of a simplified analysis carried out by DeVoe [11], assuming a thin piezoelectric layer on a thick beam, give the beam tip displacement and angle of rotation as follows [12]:

$$\delta = 3d_{zx} \frac{L^2 E_p}{t^2 E} V \tag{2.27}$$

$$\theta = 6d_{31} \frac{L E_p}{t^2 E} V \tag{2.28}$$

where corresponding to a beam of length L, thickness t and Young's modulus E_g with a piezoelectric layer having Young's modulus E_p and , an applied voltage V yields tip deflection and rotation δ and θ, respectively. For a beam of cross-sectional area A, mass density ρ, length L, and moment of inertia I, the corresponding fundamental mechanical bending resonance frequency is given by (2.29).

$$\omega_0 = 3.5160 \left(\frac{EI}{\rho L^4 A} \right)^{1/2} \tag{2.29}$$

2.1.3   Thermal Actuation

The thermal expansion of a material when its temperature is raised ΔT degrees may be exploited as an actuation mechanism [13]. Indeed, a constrained beam of length L and cross-sectional area A experiences a corresponding strain $\varepsilon = \alpha \Delta T$, where $\alpha(°C^{-1})$ is the coefficient of thermal expansion, and this causes a normal stress $\sigma = E\varepsilon$. This stress manifests either as an elongation $\delta L = EL$ if the beam is unconstrained, or as a force $F = A\sigma$ upon its constraints. The heat-driven actuator exploits this principle.

2.1.3.1 The Heat-Driven Actuator

In the heat-driven actuator (see Figure 2.8), flexures of different cross-section are configured to form a loop. Actuation, then, is achieved when a current running through the device heats a thin section A to a temperature higher than a thick section B. The resulting tip deflection, Δx, has been modeled by (2.30) [12, 13].

$$\Delta x = \frac{\alpha \Delta T L^2}{g_1 \left(0.7707 + 0.3812 \frac{t_A^2}{g_1^2} \right)}$$ (2.30)

Examination of (2.30) reveals that the deflection is independent of the layer thickness, h, and of the Young's modulus, E. The temperature increase, ΔT, however, is induced by a current I (Amps) passed through the structure, and is related to its electrical resistance, $R\,(\Omega)$, given by (2.31),

$$P_{Diss} = \frac{T - T_0}{\theta_{junct}} = I^2 R(T)$$ (2.31)

where P_{Diss} (watts) is the power dissipated, $T(^\circ C)$ is the temperature to which the structure is heated, $T_0(^\circ C)$ is the ambient temperature and θ_{junct}

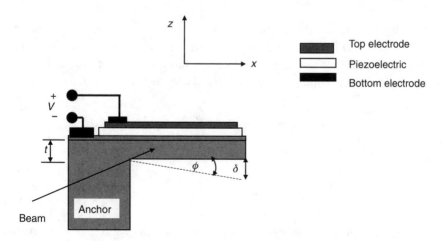

Figure 2.8 Sketch of horizontal heat-driven actuator. Typical dimensions are [13]: $L_1 + L_2 =$ 270 μm, $L_1 = 40\,\mu$m, $t_A = t_B = 4\,\mu$m, $t_C = 15\,\mu$m, $g_2 = 4\,\mu$m, $g_1 = 13.5\,\mu$m.

$(°C^{-1} - W^{-1})$ is the thermal resistance between the structure and the ambient atmosphere. For the a beam of length L and cross section as in Figure 2.8, the resistance is given by (2.32),

$$R(T) = \rho_0 \frac{L}{t_A \cdot h} \exp(\alpha_R \cdot (T - T_0)) \cong \rho_0 \frac{L}{t_A \cdot h} \cdot (1 + \alpha_R \cdot (T - T_0)) \quad (2.32)$$

where $\rho_0 (\Omega - cm)$ is the beam (conductor) resistivity, and $\alpha_R (°C^{-1})$ is the thermal coefficient of resistance of the conductor. From (2.31) and (2.32), the temperature increase is obtained as,

$$\Delta T = \frac{I^2 \cdot \rho_0 \cdot L \cdot \theta_{junct}}{h \cdot t_A - I^2 \cdot \rho_0 \cdot L \cdot \alpha_R \cdot \theta_{junct}} \quad (2.33)$$

The temperature difference between sections A and B (see Figure 2.8), due to the same current is given by,

$$\Delta T_{A-B} = I^2 \cdot \rho_0 \cdot L \cdot$$

$$\left[\frac{\theta_A}{h \cdot t_A - I^2 \cdot \rho_0 \cdot L \cdot \alpha_R \cdot \theta_A} - \frac{\theta_B}{h \cdot t_B - I^2 \cdot \rho_0 \cdot L \cdot \alpha_R \cdot \theta_B} \right] \quad (2.34)$$

where θ_A and θ_B are the thermal resistance $(°C - W^{-1})$ between sections A and B, and the ambient temperature, respectively. These parameters model the radiation of the heat conductors into the atmosphere, and are assumed to be proportional to the conductor surface area. Thus, they are modeled as,

$$\theta_A = a \cdot 2 \cdot L \cdot (h + t_A) \quad (2.35)$$

and

$$\theta_B = a \cdot 2 \cdot L \cdot (h + t_B) \quad (2.36)$$

where a is the proportionality constant that relates thermal resistance to surface area and must be determined experimentally. The final equation for temperature difference between sections is given by (2.37).

$$\Delta T_{A-B} = I^2 \cdot \rho_0 \cdot L \cdot \left[\frac{a \cdot 2 \cdot L \cdot (h + t_A)}{h \cdot t_A - I^2 \cdot \rho_0 \cdot L \cdot \alpha_R \cdot a \cdot 2 \cdot L \cdot (h + t_A)} \right.$$

(2.37)

$$\left. - \frac{a \cdot 2 \cdot L \cdot (h + t_B)}{h \cdot t_B - I^2 \cdot \rho_0 \cdot L \cdot \alpha_R \cdot a \cdot 2 \cdot L \cdot (h + t_B)} \right]$$

The deflection Δx is then calculated from (2.38) [12].

$$\Delta x = \frac{\alpha \Delta T_{A-B} L^2}{g_1 \left(0.7707 + 0.3812 \frac{t_A^2}{g_1^2} \right)}$$

(2.38)

In addition to the deflection induced by thermal heating, there exists a repulsive force between the sides of the flexure that is developed due to the fact that the current flows in opposite directions. It has been pointed out that this force prevents the two sides from coming into contact, and that it helps increase the amount of tip deflection. This extra tip deflection is given by (2.39) [13, 14],

$$\Delta x = \frac{p \cdot L^4}{27 \cdot E \cdot I}$$

(2.39)

where p (N/m^2) is the magnetic pressure in force/length between each side of the heat-drive actuator, and $I = t_A \cdot h^3 / 12$ is the beam moment of inertia.

2.1.4 Magnetic Actuation

Magnetic actuation relies on the exploitation of a magnetic field to generate a force. This magnetic force typically has one of two origins, namely, a current or a magnet. In the former case, a current I is utilized to create a magnetic flux through a magnetic circuit. The magnetic circuit has an air gap formed by the cantilever and the contact, as shown in Figure 2.9. The force across the gap is given by (2.40) [15],

$$F_M = \frac{\mu_0}{2} A \frac{(nI)^2}{d^2}$$

(2.40)

where n is the number of turns of the coil, A the area of the ferromagnetic material contact, and $d = d_c + d_0$ is the gap between the beam and the contact [14],

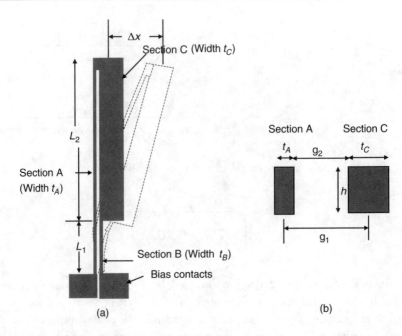

Figure 2.9 Magnetic actuation devices. (a) Current induces force across gap in magnetic circuit. (b, c) Interaction between external magnetic field and magnetized material on flap elicits torque that tends to align flap with field. The alignment occurs instantaneously for a hinged flap, (b), or gradually, (c), for a flap with an opposing resistive force. (After [16].)

with d_c representing the air-gap and d_0 representing the magnetic resistance of the magnetic circuit.

In the latter case, developed by Liu [16], an external magnetic field, H_{ext}, interacts with a permanently magnetized material of magnetization, M, disposed on a flap, [see Figure 2.9(b)], to create a torque which effects a bending of the flap by an angle θ, (2.41),

$$T_m = M w_m t_m l_m H_{ext} \cos \theta \qquad (2.41)$$

where l_m, w_m, and t_m are the length, width, and thickness of the magnetized material. This principle has been applied to the assembly of arrays of flaps [16].

2.2 Mechanical Vibrations

In Figure 2.3 of Section 2.1.2, we saw that the electrostatic load in a cantilever beam has the highest concentration at its tip. As this loading condition resembles that of the familiar springboard, we intuitively expect that the sudden application or removal of the load will lead to mechanical vibration of the beam. Indeed, a mechanical system can in general vibrate in *n* number of modes, according to the degrees of freedom it possesses [17–19]. The degrees of freedom of a mechanical system are equal to the number of parameters needed to specify its position. For example, a springboard, restrained to move up and down in the *z* direction, has one degree of freedom; a hockey puck sliding in its plane has three degrees of freedom, namely, the *x*- and *y*-displacements of the center of mass and the angle of rotation about the center of gravity; and a baseball on route to the catcher has six degrees of freedom, the *x*-, *y*-, and *z*-displacements of the center of mass and three rotations. The importance of knowing the degrees of freedom of the mechanical systems under consideration is that, in analogy with the familiar first-, and second-order RL/RC, and RLC circuits of circuit theory [20], they determine how the system responds to an excitation. In general, then, mechanical systems may be described by a first-, second-, or *nth*-order differential equation, depending on whether they are a first-, second-, or −*nth*-degree system, respectively. Unfortunately, the higher the number of degrees of freedom of a system, the more complicated it becomes to solve exactly the differential equations that describe its motion. Accordingly, systems have been categorized into few-degrees-of-freedom and many-degrees-of-freedom, with simple exact analysis techniques applied to the former and specialized approximate techniques applied to the latter. We next treat techniques for the analysis of a single-degree-of-freedom system, and introduce the Rayleigh Quotient technique, developed for analyzing a many-degrees-of-freedom system [17–19].

2.2.1 The Single-Degree-of-Freedom-System

The response of a mechanical system to an external excitation force is characterized by three parameters: its mass, M, its stiffness, denoted by the "spring constant" K, and its damping constant, D. In the simplest cases the response may be obtained as a solution to the equation of motion, given by Newton's second law, Force = mass × acceleration. In the case of our cantilever beam with possible displacement in the *z* direction and neglecting gravity, if an external force F $\sin\omega t$ is applied, the equation of motion takes the form (2.42).

$$M\ddot{z} + D\dot{z} + Kz = F \sin \omega t \tag{2.42}$$

where the dot denotes derivative with respect to time. This equation is known as the differential equation of motion of a single-degree-of-freedom system [17]. It states that the motion *z* elicited by an applied force, $F\sin\omega t$, is the result of the balance between the applied force, on the one hand, and the resultant of the impulsive motion, $M\ddot{z}$, the viscous damping force experienced by the system once it starts to develop a speed, $D\dot{z}$, and the "spring" force Kz which commands the system to return to its equilibrium position. This equation is analogous to that of Kirchhoff's Voltage Law for a series RLC circuit [20]. The four terms in (2.42) appropriately denote the inertia force, the damping force, the spring force, and the external force, respectively. Equation (2.42) is a second-order linear differential equation with familiar solutions [20]. The system can be further simplified by adopting the idealization that there is no viscous damping. The solution then is given by

$$M\ddot{z} + Kz = 0 \qquad (2.43)$$

or

$$\ddot{z} = -\frac{K}{M} z \qquad (2.44)$$

which has a general solution

$$z = A \sin t \sqrt{\frac{K}{M}} + B \cos t \sqrt{\frac{K}{M}} \qquad (2.45)$$

where A and B are arbitrary constants. This represents temporal evolution of the beam's spatial shape under undamped conditions, one cycle of which occurs when $\omega_n = t\sqrt{K/M}$, the so-called "natural circular frequency," varies through 360° or 2π radians [16]. The natural angular frequency f_n, is

$$f_n = \frac{\omega_n}{2\pi} = \frac{1}{2\pi} \sqrt{\frac{K}{M}} \qquad (2.46)$$

and is measured in cycles per second (Hz). Physically, this frequency embodies the fact that, left to itself, a system possessing mass, M, and spring constant, K, will vibrate in response to the force exerted by its own mass. Since no external force is required in order to excite these vibrations, they are referred to as free vibrations. Implications of this frequency, in the context of microwave applications, will be discussed in Chapter 3, in what pertains to the maximum switching frequency of the cantilever beam-type switch.

2.2.2 The Many-Degrees-of-Freedom System

For systems of many degrees of freedom (e.g., the interdigitated comb-drive capacitor), a frequency determination of the vibration modes from the differential equation often becomes so complicated as to be practically impossible. In such cases a generalized energy method, known as the method of Rayleigh, is necessary [17–19]. The method may be demonstrated with reference to the cantilever beam example above.

If we start with the assumption that the motion is harmonic, the frequency can be calculated in a very simple manner from an energy consideration [17–19]. As is learned in a physics course on basic mechanics [21], the kinetic and potential (elastic) energies of a mechanical system oscillate in such a way that in the middle of the motion, when the system passes through its equilibrium configuration, the kinetic energy is a maximum and the potential energy a minimum, but when either extreme position is reached, there is zero kinetic energy and maximum of potential energy. Analogously, in the middle of a swing the beam has considerable kinetic energy, whereas in either extreme position it stands still momentarily and has no kinetic energy left. In this state the beam is storing all the energy as elastic tension. At any position between the middle and the extreme, both elastic and kinetic energy are present, adding in fact to a constant sum if the system is isolated. By equating the kinetic energy in the middle of a vibration to the elastic energy in an extreme position, a relationship is obtained for the resonance frequency. These energies are calculated as follows.

When the beam reaches the peak of its displacement z_0, the potential or elastic energy due to the spring force is $\int_0^z Kz \cdot dz = \frac{1}{2}Kz^2$. At any instant of time the kinetic energy is $1/2 \ mv^2$. If we assume that the motion is given by $z = z_0 \sin\omega t$, then the corresponding velocity is $v = z_0\omega \cos \omega t$. On the other hand, the potential energy at the peak displacement is $1/2 \ Kz_0^2$, and the kinetic energy in the equilibrium position, where the velocity is maximum, is $1/2 \ mv_{max}^2 = 1/2 \ m\omega^2 z_0^2$.

Therefore, equating energies we have

$$\frac{1}{2} Kz_0^2 = \frac{1}{2} m\omega^2 z_0^2 \tag{2.47}$$

from which $\omega^2 = K/M$, independent of the amplitude z_0. In this way, the lowest or fundamental frequency is given by the so-called Rayleigh's Quotient, (2.48).

$$\omega^2 = \frac{\frac{1}{2} Kz_0^2}{\frac{1}{2} mz_0^2} = \frac{V_{max}}{T_{max}} \tag{2.48}$$

2.2.3 Rayleigh's Method

As shown above, the fundamental frequency is given by the ratio of the maximum potential energy to the maximum kinetic energy of the system (i.e., Rayleigh's Quotient). A method developed by Rayleigh generalizes the above procedure to address the problem of finding the lowest or fundamental frequency of vibration for a general system possessing distributed mass and distributed flexibility [17–19]. The governing equation of the structure is given by

$$EI \frac{\partial^4 u(z,t)}{\partial z^4} = -\mu_1 \frac{\partial^2 u(z,t)}{\partial t^2} \tag{2.49}$$

where E is the modulus of elasticity, I is the moment of inertia, μ_1 is the mass per unit length, $u(z, t)$ is the geometrical deformation of the structure under vibration, and t is time. The method is based on assuming the shape for the first normal elastic curve for the mechanical deformation corresponding to the maximum displacement of the vibration. Let the geometrical deformation of the structure under vibration, $u(z, t)$, be separable into a product of independent spatial and temporal functions, $U(z)$ and $f(t)$, respectively. Then, with the continuum vibration given by [22]

$$u(z,t) = U(z)f(t) \tag{2.50}$$

a continuum velocity given by

$$\dot{u}(z,t) = U(z)\dot{f}(t) \tag{2.51}$$

and a continuum potential given by

$$\frac{\partial u}{\partial z}(z,t) = \frac{\partial U(z)}{\partial z} f(t) \tag{2.52}$$

substituting (2.50), (2.51), and (2.52) into (2.49), we obtain

$$EI \frac{\partial^4 U(z)}{\partial z^4} = \mu_1 \omega^2 U(z) \tag{2.53}$$

The corresponding potential energy V, and kinetic energy T, are given by

$$V = \frac{1}{2} \int_0^L EI \left[\frac{\partial U(z)}{\partial z} f(t) \right]^2 dz \tag{2.54}$$

and

$$T = \frac{1}{2} \int_0^L M(z) \left[U(z) \dot{f}(t) \right]^2 dz \qquad (2.55)$$

where $M(z)$ is the distributed mass.

The Rayleigh's quotient is then given by

$$\omega^2 = \frac{\int_0^L EI \left[\dfrac{\partial U(z)}{\partial z} \right]^2 dz}{\int_0^L M(z) \left[U(z) \right]^2 dz} \qquad (2.56)$$

As an example [17] of the application of this procedure, we calculate the fundamental frequency of a cantilever beam, as shown in Figure 2.10. We take the spatial profile of the continuum vibration to be given by the curve

$$z = z_0 \left(1 - \cos \frac{\pi x}{2L} \right) \qquad (2.57)$$

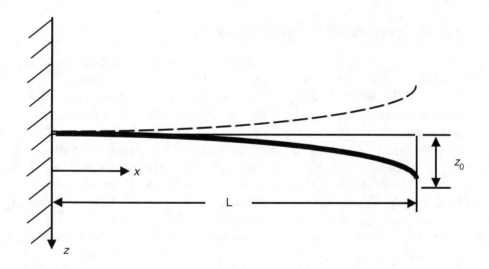

Figure 2-10 Quarter cosine wave as a Rayleigh shape for a cantilever. (*After:* [17].)

Substituting (2.57) into (2.56), we obtain

$$\omega^2 = \frac{\int_0^L EI \left[\frac{\partial U(z)}{\partial z} \right]^2 dz}{\int_0^L M(z) [U(z)]^2 dz} = \frac{\dfrac{\pi^4}{64} \dfrac{EI}{L^3} z_0^2}{\mu_1 z_0^2 L \left(\dfrac{3}{4} - \dfrac{2}{\pi} \right)} \tag{2.58}$$

or

$$\omega = \frac{\pi^2}{8 \sqrt{\dfrac{3}{4} - \dfrac{2}{\pi}}} \sqrt{\frac{EI}{\mu_1 L^4}} = \frac{3.66}{L^2} \sqrt{\frac{EI}{\mu_1}} \tag{2.59}$$

where μ_1 is the mass density per unit length.

Rayleigh's method is an approximate way of estimating the frequency of the first mode of vibration of a system, based on an assumed motion. However, if the assumed motion happens to be exact, then Rayleigh's method gives the exact solution to the natural frequency. This is evident from (2.29). By assuming a constant mass, M, and the motion $z = z_0 \sin\omega t$, Rayleigh's quotient gives $\omega^2 = K/M$, which is identical to (2.46).

2.3 Computer-Aided Design of MEMS

The usual arguments that justify devoting resources to the development of tools that aid in the analysis and design of integrated circuits are also applicable when it comes to MEMS. Indeed, MEMS also can benefit from a reduced number of fabrication iterations and experimentation, and from the attainment of at least second-pass success. Contrary to the situation in IC-CAD tools, however, CAD tools for MEMS are a long way from reaching maturity. This state of affairs is rooted in a number of peculiar facts germane to MEMS. Since there is a wide variety of potential MEM devices, as derived from the varied actuation mechanisms they exploit, efforts must be addressed at simulating *classes* of devices (e.g., electrostatic, magnetic, thermoelectromagnetic). Also there is the fact that the simulation of MEM devices requires the integration of rather disparate disciplines and paradigms, (e.g., materials and fabrication process modeling, solid modeling, mechanical simulation, physical device prototyping and analysis, and compact lumped-parameter models for behavioral device modeling in dynamic system simulation). In the third place we have the rich behavior and operating regimes displayed by MEM devices (e.g., energy-conserving or dissipative, quasi-static or dynamic, linear and/or nonlinear), which makes the numerical formulation and solution of the problem rather nontrivial. Finally, we have the fact that the fabrication technology itself is in a state of flux, which means that materials and processes are constantly changing.

In this section we present a review of the fundamental elements of CAD for MEMS, namely, materials and fabrication process modeling, solid modeling, numerical simulation, and lumped-parameter modeling. A detailed treatment of actual numerical methods, however, is clearly beyond the scope of this book.

2.3.1   MEMS Materials and Fabrication Process Modeling

As described in Chapter 1, MEMS are three-dimensional (3D) structures. Therefore, the first step in their simulation is to describe the 3D structures in a way that reflects, as close as possible, the effects of the fabrication processes employed (e.g. diffusion, deposition, and etching, on device geometry and material composition). The standard way of accomplishing this task in conventional IC process modeling is via process simulator programs like SUPREM [23] and SAMPLE [24, 25], which, through numerical simulation of the physical processes, yield the 2D device cross section of the cumulative process sequence steps. In the case of MEMS, however, since it is necessary to generate the resulting 3D geometries of the standing structures, the standard tools are merged with solid modeling programs [26–29]. In a different approach, instead of numerically simulating the details of the physical fabrication process steps, a direct geometrical description of their effects is utilized to *emulate* the final structure [30].

2.3.2   Solid Modeling

Once the 3D geometry and material composition of the MEM structures have been defined, it is necessary to prepare the spatial environment, which contains them for mechanical and electrical numerical simulation. This entails transforming the 3D geometries of the devices into finite-element meshes amenable to the application of pertinent boundary conditions [28].

2.3.3   Numerical Simulation of MEMS

Numerical simulation programs aim at achieving a "first principles" representation of the MEM device behavior through a set of coupled mechanical/electrical partial differential equations, utilizing detailed materials composition, fabrication process, and layout/geometry as point of departure. In general, the mechanical response of the MEM device is caused by a force (e.g., electrical or magnetic), and manifests itself on the device as a change in stress level and displacement.

2.3.3.1 Mechanical Simulation

The goal of the mechanical simulation is to determine the elastic deformation of a three-dimensional structure in response to applied forces. This deformation is obtained from a solution of the fundamental equations for describing the behavior of a solid [31]:

1. The conditions of equilibrium:

$$\frac{\partial \sigma_x}{\partial x} + \frac{\partial \tau_{yx}}{\partial y} + \frac{\partial \tau_{zx}}{\partial z} + \bar{p}_{V_x} = 0, \quad \tau_{xy} = \tau_{yx}$$

$$\frac{\partial \tau_{xy}}{\partial x} + \frac{\partial \sigma_y}{\partial y} + \frac{\partial \tau_{zy}}{\partial z} + \bar{p}_{V_y} = 0, \quad \tau_{xz} = \tau_{zx} \qquad (2.60)$$

$$\frac{\partial \tau_{xz}}{\partial x} + \frac{\partial \tau_{yz}}{\partial y} + \frac{\partial \sigma_z}{\partial z} + \bar{p}_{V_z} = 0, \quad \tau_{yz} = \tau_{zy}$$

where σ_i, τ_{ij}, and $\bar{p}_{V_i}$ are the stress components in the direction i, the shear stresses in the plane ij, and the force per unit area in the direction i, respectively.

2. The material laws:

$$\begin{bmatrix} \varepsilon_x \\ \varepsilon_y \\ \varepsilon_x \\ \gamma_{xz} \\ \gamma_{xz} \\ \gamma_{yz} \end{bmatrix} = \frac{1}{E} \begin{bmatrix} 1 & -v & -v & 0 & 0 & 0 \\ -v & 1 & -v & 0 & 0 & 0 \\ -v & -v & 1 & 0 & 0 & 0 \\ 0 & 0 & 0 & 2(1+v) & 0 & 0 \\ 0 & 0 & 0 & 0 & 2(1+v) & 0 \\ 0 & 0 & 0 & 0 & 0 & 2(1+v) \end{bmatrix} \begin{bmatrix} \sigma_x \\ \sigma_y \\ \sigma_z \\ \tau_{xy} \\ \tau_{xz} \\ \tau_{yz} \end{bmatrix} \qquad (2.61)$$

where E is Young's modulus or the modulus of elasticity, v is *Poisson's* ratio, and $\gamma_{ij} = \tau_{ij}/G$, where G is the *shear modulus of elasticity*. The material laws are usually called Hooke's law when dealing within the linear, elastic range of a homogeneous solid [30].

3. The strain-displacement relations:

$$\varepsilon_x = \frac{\partial u}{\partial x} + \frac{1}{2}\left[\left(\frac{\partial u}{\partial x}\right)^2 + \left(\frac{\partial v}{\partial x}\right)^2 + \left(\frac{\partial w}{\partial x}\right)^2\right].$$

$$\varepsilon_y = \frac{\partial u}{\partial y} + \frac{1}{2}\left[\left(\frac{\partial u}{\partial y}\right)^2 + \left(\frac{\partial v}{\partial y}\right)^2 + \left(\frac{\partial w}{\partial y}\right)^2\right]$$

$$\varepsilon_z = \frac{\partial u}{\partial z} + \frac{1}{2}\left[\left(\frac{\partial u}{\partial z}\right)^2 + \left(\frac{\partial v}{\partial z}\right)^2 + \left(\frac{\partial w}{\partial z}\right)^2\right]$$

$$\varepsilon_{xy} = \frac{1}{2}\left(\frac{\partial v}{\partial x} + \frac{\partial u}{\partial y} + \frac{\partial u}{\partial x}\frac{\partial u}{\partial y} + \frac{\partial v}{\partial x}\frac{\partial v}{\partial y} + \frac{\partial w}{\partial x}\frac{\partial w}{\partial y}\right)$$

$$\varepsilon_{xz} = \frac{1}{2}\left(\frac{\partial w}{\partial x} + \frac{\partial u}{\partial z} + \frac{\partial u}{\partial x}\frac{\partial u}{\partial z} + \frac{\partial v}{\partial x}\frac{\partial v}{\partial z} + \frac{\partial w}{\partial x}\frac{\partial w}{\partial z}\right)$$

$$\varepsilon_{yz} = \frac{1}{2}\left(\frac{\partial w}{\partial y} + \frac{\partial v}{\partial z} + \frac{\partial u}{\partial y}\frac{\partial u}{\partial z} + \frac{\partial v}{\partial y}\frac{\partial v}{\partial z} + \frac{\partial w}{\partial y}\frac{\partial w}{\partial z}\right)$$

$$(2.62)$$

where u, v, and w, the three components of the displacement vector at a point in a solid, are mutually orthogonal. The corresponding shear strains are given by $\gamma_{ik} = 2\varepsilon_{ik}$ $(i \neq k)$ [31].

4. The boundary conditions:

In general, the surface of a solid is considered to be made up of two complementary surfaces Su and Sp [31], where Su encompasses structure regions whose displacement is known, whereas Sp encompasses everything else, and in particular, the surface where the external forces are applied. Surface forces per unit area applied externally are referred to as *tractions* [31]. For electromechanical structures, whose surfaces are normally assumed to be perfect conductors, the applied electrostatic pressure load gives the traction,

$$P = \frac{\sigma^2}{2\varepsilon} \tag{2.63}$$

where σ is the surface charge density on the metallic structures [32]. Rather than developing new programs to perform the mechanical simulation, a common practice is to take advantage of mature finite-element modeling tools, developed in the mechanical engineering field, such as ABAQUS [33] and ANSYS [34].

2.3.3.2   Electrostatic Simulation

The application of an electric field or voltage to a MEMS structure elicits electrostatic forces, which, through (2.46), generate tractions on their surfaces, thus causing them to deform. Since the MEMS structures consist of conductors, dielectrics, and coatings, the induced deformation changes the capacitance of the system, which in turn changes the forces on them [27]. This means that a self-consistent electromechanical analysis is needed to simulate the device. The heart of the electrostatic simulation is the computation of the surface charges on the conductors of the system. This invariably entails numerically solving the integral form of Laplace's equation,

$$\psi(x) = \int_{Surfaces} G(x, x')\sigma(x')dS' \qquad (2.64)$$

by a variety of numerical techniques [35–38], where as above, σ is the surface charge density, dS' is the incremental surface area, ψ is the applied surface potential, and $G(x, x')$

$$G(x, x') = \frac{1}{4\pi\varepsilon\|x - x'\|} \qquad (2.65)$$

is the Green's function.

2.3.3.3   MEMS Simulation Systems

MEMS simulation systems typically integrate tools for materials and fabrication process modeling, solid modeling, and mechanical simulation. In addition, depending on the class of MEMS devices or phenomena targeted, pertinent modules are included to tackle such issues as electrostatic simulation [32], coupling to fluid dynamics phenomena [39], behavior involving magnetic materials [40], thermal behavior [41], and electrothermomechanical behavior [42]. Two well known examples of commercialized MEMS simulation systems are the SOLIDIS system [42], marketed by ISE [43] and aimed at devices governed by electrostatic forces as well as thermomechanical and piezoelectric interactions and versions related to the MEMCAD system [44], marketed as CoventorWare by Coventor, Inc. [45]; and as IntelliCAD by IntelliSense [46], aimed at electrostatic-actuation-based MEMS devices.

While numerical device simulation enables the detailed analysis and visualization of the inner workings of a device, it is usually accompanied by two drawbacks; namely, long simulation times and an intractably large amount of information, both of which make its use impractical for system design. This is why major CAD tools now include lumped-element simulation facilities, where the mechanical system is modeled via a simplified reduced-order model containing only lumped elements, in particular, masses, dampers, and springs [47–50].

2.4 Exercises

Exercise 2.1 Capacitance Magnitude
A parallel-plate structure has plate areas of $A = 10\ \mu\text{m} \times 50\ \mu\text{m}$, interplate distance of $d = 2\ \mu\text{m}$, and air ($\varepsilon_r = 1.0$) as dielectric. Calculate its capacitance.

Solution:
Substituting into (2.1), one obtains:

$$C = \frac{\left(8.85418 \cdot 10^{-12}\ \dfrac{F}{m}\right) \cdot \left(10 \cdot 10^{-6}\ m \times 50 \cdot 10^{-6}\ m\right)}{2 \cdot 10^{-6}\ m}$$

$$= 2.214 \cdot 10^{-15}\ F = 2.214\text{fF}$$

Exercise 2.2 Energy Stored in Capacitor
The capacitor of Exercise 2.1 sustains a potential difference of 10V. Calculate the energy stored in its electric field.

Solution:
The units of capacitance is the farad $= \dfrac{\text{coul}^2}{N-m}$, and that of potential the volt $= \dfrac{J}{\text{coul}}$, where the unit of energy is the joule $= J = N-m$. Therefore, using these units and substituting into (2.2), one obtains:

$$U = \frac{1}{2} \cdot \left(2.214 \cdot 10^{-15}\ \frac{\text{coul}^2}{N-m}\right) \cdot \left(10\ \frac{J}{\text{coul}}\right)^2 = 1.107 \cdot 10^{-13}\ \frac{J^2}{N-m}\ 110.7\text{fJ}$$

Exercise 2.3   Spring Constant Units

A spring stretched a distance Δx from equilibrium experiences a restoring force proportional to this distance, $F = -K\Delta x$. This is the well-known Hooke's law, where the proportionality constant K is called the *spring constant*. Find the units of K in the International System (SI).

Solution:

In SI units, the force F is given in newtons (N), and the displacement in meters (m). Therefore, the spring constant has units of

$$[K] = \frac{[F]}{[\Delta x]} = \frac{N}{m}$$

Exercise 2.4   Threshold Voltage for Electrostatic Closing Deflection of Beam

The cantilever beam of Figure 2.2 has the following beam-to-substrate distance (d), length (l), width (b), thickness (t), and Young's modulus (E) given by $d = 2\,\mu m$, $l = 25\,\mu m$, $b = 25\,\mu m$, $t = 0.2\,\mu m$, and $E = 0.72 \cdot 10^{12}\,\text{dyn/cm}^2 = 7.2\ 10^{10}$ N/m^2, respectively. Calculate its defection threshold voltage.

Solution:

Substituting into (2.15), one obtains:

$$V_{th} = \sqrt{\frac{18 \cdot \left(7.2 \cdot 10^{10}\,\frac{N}{m}\right) \cdot \left[25 \cdot 10^{-6}\,m\left(0.2 \cdot 10^{-6}\,m\right)^3 \Big/ 12\right]\left(2 \cdot 10^{-6}\,m\right)^3}{5 \cdot \left(8.8541 \cdot 10^{-12}\,\frac{F}{m}\right) \cdot \left(25 \cdot 10^{-6}\,m\right)^4 \cdot \left(25 \cdot 10^{-6}\,m\right)}}$$

$$\approx 20\ \text{volts}$$

where the beams moment of inertia was assumed to be given by $I = bt^3/12$.

Exercise 2.5   Threshold Voltage for Electrostatic Opening of Deflected Beam

Assume that, when deflected, the beam in Exercise 2.4 has an effective beam-to-substrate distance $d_c = d/4$. Calculate by how much must the closing actuation voltage be reduced for the beam to begin springing back to its undeflected state.

Solution:

Substituting in (2.17), one obtains:

$$V_{th-open} = \left(2 \cdot 10^{-6}\,\text{m} - \frac{2 \cdot 10^{-6}\,\text{m}}{4}\right) \cdot$$

$$\sqrt{\frac{2 \cdot \left[\dfrac{\left(25 \cdot 10^{-6}\,\text{m}\right) \cdot \left(0.2 \cdot 10^{-6}\,\text{m}\right)^3 \left(7.2 \cdot 10^{10}\,\dfrac{\text{N}}{\text{m}^2}\right)}{4 \cdot \left(25 \cdot 10^{-6}\,\text{m}\right)^3}\right] \cdot \dfrac{\left(2 \cdot 10^{-6}\,\text{m}\right)}{4}}{3 \cdot \left(8.85418 \cdot 10^{-12}\,\dfrac{\text{F}}{\text{m}}\right) \cdot \left(25 \cdot 10^{-6}\,\text{m} \times 25 \cdot 10^{-6}\,\text{m}\right)}}$$

$$= 7.9 \text{ volts}$$

Thus, the closing actuation voltage must be reduced by

$$V_{reduction} = V_{th-close} - V_{th-open} = 20 - 7.9 = 12.1 \text{ volts}$$

Exercise 2.6   Cantilever Beam Resonance Frequency

The beam described above has a mass density per unit length of $\mu = 1.542 \cdot 10^8$ kg. Calculate its first resonance frequency.

Solution:

Substituting in (2.40), one obtains:

$$f_{res} = \frac{3.66}{2\pi \cdot \left(25 \cdot 10^{-6}\,\text{m}\right)^2}$$

$$\sqrt{\frac{\left(7.2 \cdot 10^{10}\,\text{N/m}\right) \cdot \left[25 \cdot 10^{-6}\,\text{m}\left(0.2 \cdot 10^{-6}\,\text{m}\right)^3 \Big/ 12\right]}{1.542 \cdot 10^{-8}\,\text{kg}}} = 260 \ kHz$$

2.5   Summary

In this chapter we have presented an introduction to the physics of various actuation mechanisms, namely, electrostatic, piezoelectric, thermal, and magnetic, and some fundamental devices that exploit them. In the first place, we dealt with the electrostatic parallel-plate capacitor actuator, and the interdigitated comb-drive capacitor actuator. While the former is a device of simple construction and operation, it possesses the drawback that its force drops as $1/x^2$ with interelectrode distance. The latter, on the other hand, provides an actuation force that is independent of displacement distance, x, but requires careful implementation to achieve strictly one-dimensional actuation. In the second place, we introduced the piezoelectric cantilever actuator. We saw that its displacement varies linearly and continuously with the applied voltage, so that displacement may be precisely controlled. In the third place, we introduced the heat-driven actuation. We saw that stress generated by the propensity of a material to thermally expand upon being heated may be exploited to exert a force when it is constrained. Further, the differential in temperature rise of two beams of appropriate dimensions attached to each other may be used as an actuator. Finally, we introduced two types of magnetic actuators. The first, which was current-driven, derived its force from the magnetic flux induced in a magnetic circuit containing a gap. Similar to the electrostatic case, the force is inversely proportional to the gap, but it is proportional to the square of the number of turns and the current through the coil generating the magnetic field. When a cantilever is part of the magnetic circuit, the force across the gap causes the beam to move such as to close the gap. The second method involved the torque induced on a flap on which a magnetic material was disposed when an external magnetic field was turned on.

Under the influence of their own mass, mechanical structures undergo vibrations. We presented methods for calculating the fundamental frequency of vibration, namely, one suitable for simple single-degree-of-freedom structures, and the more general Rayleigh's energy method, appropriate for structures possessing distributed mass and distributed flexibility. We have also summarized the fundamentals of CAD for MEMS. In particular, we presented the elements of "first principles" numerical modeling, which is appropriate to gain a deep understanding of the inner workings of a device, as well as to expedite device development.

The performance of electrostatic, thermal, and magnetic actuators is characterized by a number of parameters, such as power consumption, speed, force, displacement range/stroke distance, and size. When considering their application as microwave circuit devices, three of these parameters are of particular interest: power consumption, speed, and size. In Chapters 3 and 4, we will deal with the design considerations for the practical application of the most fundamental electrostatic MEM device, the cantilever beam, in its two main microwave applications, namely, those of switch and of resonator. We will address its operation and specifications, as well as important microwave, material, and mechanical considerations. We will also discuss pertinent circuit models and factors affecting its ultimate performance.

References

[1]   Comtois, J. H., "Structures and Techniques for Implementing and Packaging Complex, Large Scale Microelectromechanical Systems Using Foundry Fabrication," Ph.D. dissertation, Air Force Institute of Technology, 1996.

[2]   Sniegowski, J. J., and E. J. Garcia, "Microfabricated Actuators and Their Application to Optics," *Proc. SPIE—Int. Soc. Opt. Eng.* (USA), Vol. 2383, San Jose, CA, February 7–9, 1995, pp. 46–64.

[3]   Price, R. H., J. E. Wood, and S. C. Jacobsen, "The Modeling of Electrostatic Forces in Small Electrostatic Actuators," *Tech. Digest. IEEE Solid-State Sensor and Actuator Workshop*, June 1988, pp. 131–135.

[4]   Trimmer, W. S. N., and K. J. Gabriel, "Design Considerations for a Practical Electrostatic Micromotor," *Sensors and Actuators*, Vol. 11, No. 2, 1987, pp. 189–206.

[5]   Peterson, K. E., "Dynamic Micromechanics on Silicon: Techniques and Devices," *IEEE Trans. Electr. Dev.,* Vol. ED-25, 1978, pp. 1242–1249.

[6]   Zavracky, P. M., S. Majumder, and N. E. McGruer, "Micromechanical Switches Fabricated Using Nickel Surface Micromachining," *J. Microelectromechanical Systems*, Vol. 6, 1997, pp. 3–9.

[7]   Tang, W. C., T-C. H. Nguyen, and R. T. Howe, "Laterally Driven Polysilicon Resonant Microstructures," *Proc. IEEE Microelectromechanical Systems*, 1989, pp. 53–59.

[8]   T. Ikeda, *Fundamentals of Piezoelectricity*, New York: Oxford Science Publications, 1996.

[9]   Muralt, P., "MEMS: A Playground for New Thin Film Materials," Available on-line at http://semiconductors.unaxis.comenchiponline_72dpiissue617-19.pdf.

[10] Weinberg, M. S., "Working Equations for Piezoelectric Actuators and Sensors," *ASME/IEEE J. Microelectromechanical Systems*, Vol. 8, No. 4, December 1999, pp. 529–533.

[11] DeVoe, D., "Thin Film Zinc Oxide Microsensors and Microactuators," Ph.D. dissertation, Department of Mechanical Engineering, University of California, Berkeley, CA, 1997.

[12] Cheng, H.-M., et al., "Modeling and Control of Piezoelectric Cantilever Beam Micromirror and Microlaser Arrays to Reduce Image Banding in Electrophotographic Processes," *J. Micromech. Microeng.* Vol. 11, 2001, pp. 1–12.

[13] Philips, M. W., "Design and Development of Microswitches for Microelectromechanical Relay Matrices," M.S.E.E. Thesis, Air Force Institute of Technology, June 1995.

[14] Guckel, H., et al., "Therm-magnetic Metal Flecure Actuators," *Proc. Transducers Conf.*, San Francisco, CA, June 1992, pp. 73–75.

[15] Schlaak, H. F., "Potentials and Limits of Microelectromechanical Systems for Relays and Switches," *21st International Conference on Electrical Contacts*, Zurich, Switzerland, September 9–12, 2002, pp. 19–30.

[16]   Yi, Y., and Chang Liu, "Magnetic Actuation of Hinged Microstructures," *IEEE J. Microelectromechanical Syst.*, Vol. 8, No. 1, March 1999, pp. 10–17.

[17]   Den Hartog, J. P., *Mechanical Vibrations*, Third Edition, New York and London: McGraw-Hill, 1947.

[18]   Skudrzyk, E., *Simple and Complex Vibratory Systems*, University Park, PA: The Pennsylvania State University Press, 1968.

[19]   Waver, Jr., W., S. P. Timoshenko, and D. H. Young, *Vibration Problems in Engineering*, Fifth Edition, New York: John Wiley & Sons, Inc., 1990.

[20]   Desoer, C. A., and E. S. Kuh, *Basic Circuit Theory*, New York: McGraw-Hill, 1969.

[21]   Resnick, R., and D. Halliday, *Physics, Part I*, New York: John Wiley & Sons, Inc., 1966.

[22]   Boser, B. E., R. T. Howe, and A. P. Pisano, UC Berkeley Extension Course Notes: Monolithic Surface-Micromachined Inertial Sensors: Design of Closed-Loop Integrated Accelerometers and Rate Gyroscopes, May 23–24, 1995.

[23]   Ho., C. P., et al., "VLSI Process Modeling SUPREM-III," *IEEE Trans. Electron Dev.*, Vol. ED-30, No. 11, November 1983, pp. 1438–1452.

[24]   Oldham, W. G., et al., "A General Simulator for VLSI Lithography and Etching Processes: Part I—Application to Projection Lithography," *IEEE Trans. Electron Dev.*, Vol. ED-26, No. 4, April 1979, pp. 717–722.

[25]   Oldham, W. G., et al., "A General Simulator for VLSI Lithography and Etching Processes: Part II—Application to Deposition and Etching," *IEEE Trans. Electron Dev.*, Vol. ED-27, No. 8, August 1980, pp. 1455–1559.

[26]   Koppelman, G. M., "OYSTER, A Three-Dimensional Structural Simulator for Microelectromechanical Design," *Sensors and Actuators*, Vol. 20, 1989, pp. 179–185.

[27]   Maseeh, F., R. M. Harris, and S. D. Senturia, "A CAD Architecture for Microelectromechanical Systems," *IEEE Microelectromechanical Systems Workshop*, 1990, pp. 44–49.

[28]   Buser, R. A., and N. F. de Rooij, "CAD for Silicon Anisotropic Etching," *IEEE Microelectromechanical Systems Workshop*, 1990, pp. 111–112.

[29]   Crary, S., and Y. Zhang, "CAEMEMS: An Integrated Computer-Aided Engineering Workbench for Microelectromechanical Systems," *IEEE Microelectromechanical Systems Workshop*, 1990, pp. 113–114.

[30]   Ostenberg, P. M., and Senturia, S. D., " 'MEMBUILDER': An Automated 3D Solid Model Construction Program for Microelectromechanical Structures," *The 8th Int. Conf. Solid-State Sensors and Actuators, and Eurosensors IX*, Stockholm, Sweden, June 25–29, 1995, pp. 21–24.

[31]   Pilkey, W. D., and W. Wunderlich, *Mechanics of Structures: Variational and Computational Methods*, Boca Raton, FL: CRC Press, 1994.

[32]   Ostenberg, P., et al., "Self-Consistent Simulation and Modeling of Electrostatically Deformed Diaphragms," *IEEE Microelectromechanical Systems Workshop*, 1994, pp. 28–32.

[33]   Hibbit, Karlsson, and Sorensen, Inc., Providence, RI.

[34] ANSYS Users Manual, Revision 5.1, Swanson Analysis Systems Inc., Houston, PA, 1994.

[35] Shi. F., P. Ramesh, and S. Mukherjee, "Simulation Methods for Microelectromechanical Structures (MEMS) with Application to a Microtweezer," *Computers & Structures*, Vol. 56, No. 5, 1995, pp. 769–783.

[36] Nabors, K., and J. White, "FastCap: A Multipole-Accelerated 3D Extraction Program," *IEEE Trans. on Computer-Aided Design*, Vol. 10, 1991, pp. 1447–1459.

[37] Cai X., et al., "Self-Consistent Electromechanical Analysis of Complex 3D Microelectromechanical Structures Using Relaxation/Multipole-Accelerated Method," *Sensors and Materials*, Vol. 6, No. 2, 1994, pp. 85–99.

[38] Senturia, S. D., N. Aluru, and J. White, "Simulating the Behavior of MEMS Devices: Computational Methods and Needs," *IEEE Computational Science & Engineering*, 1997, pp. 30–43.

[39] Sakamoto, Y., "Dynamic Simulation Method of MEMS by Coupling Electrostatic Field, Fluid Dynamics, and Membrane Deflection," http://plast.mech.kyoto-u.ac.jp/sakamoto/research/research.html.

[40] Schneider, M., J. G. Korvink, and H. Baltes, "Magnetostatic Modeling of an Integrated Microconcentrator," *The 8th Int. Conf. Solid-State Sensors and Actuators, and Eurosensors IX*, Stockholm, Sweden, June 9–12, 1995, pp. 5–8.

[41] Jaeggi, D., et al., "Overall System Analysis of a CMOS Thermal Converter," *The 8th Int. Conf. Solid-State Sensors and Actuators, and Eurosensors IX*, Stockholm, Sweden, June 9–12, 1995, pp. 112–115.

[42] Funk, J. M., et al., "SOLIDIS: A Tool for Microactuator Simulation in 3D," *J. Microelectromechanical Syst.*, Vol. 6, No. 1, 1997, pp. 70–82.

[43] ISE Integrated Systems Engineering AG, Technopark, Techparkstrasse 1, CH-8005, Zurich, Switzerland.

[44] Senturia, S. D., CAD for Microelectromechanical Systems," *The 8th Int. Conf. Solid-State Sensors and Actuators, and Eurosensors IX*, Stockholm, Sweden, June 9–12, 1995, pp. 5–8.

[45] Microcosm Technologies, 101 Rogers Street, Cambridge, MA, 02142.

[46] IntelliSense Corporation, 16 Upton Drive, Wilmington, MA, 01887.

[47] Pelz, G., et al., "Simulating Microelectromechanical Systems," *IEEE Circuits & Devices Magazine*, 1995, pp. 10–13.

[48] Senturia, S. D., "CAD Challenges for Microsensors, Microactuators and Microsystems," *IEEE Proc.*, Vol. 86, No. 8, 1998, pp. 1611–1126.

[49] Tilmans, H. A. C., "Equivalent Circuit Representation of Electromechanical Transducers: I. Lumped-Parameter Systems," *J. Micromech. Microeng,*. Vol. 6, No. 2, 1996, pp. 157–176.

[50] Crandall, S. H., et al., *Dynamics of Mechanical and Electromechanical Systems*, New York: McGraw-Hill, 1968.

3

Fundamental MEMS Devices: The MEM Switch

3.1 Introduction

The first MEMS device developed with the purpose of engineering a microwave signal processing function was the cantilever beam [1, 2]. While in the last chapter we concerned ourselves with a purely theoretical/physical view of the cantilever beam, in this chapter we deal in detail with the fundamental device design and circuit aspects of one of its two main microwave applications, namely, that of a switch for signal routing applications. In particular, we treat the microwave, material, mechanical, power handling and reliability considerations that enter into its design and application in radio frequency and microwave systems. We will also discuss pertinent circuit models and factors affecting its fundamental performance parameters, including isolation, insertion loss, actuation voltage, and switching speed. In the next chapter, we will address the other key application of the cantilever beam, namely, that of resonator for oscillator and filtering applications.

3.2 The Cantilever Beam MEM Switch

Because of their simplicity, the most common and the most basic electrostatically actuated surface micromachined MEMS are the cantilever and the doubly supported beams [3, 4], as shown in Figure 3.1. Consequently, they embody the basic building blocks to which all electrostatically actuated MEM microwave switches may be traced. An early cantilever beam-type MEM *switch* was first developed by Petersen [1]. It had the form of a cantilever beam composed of a thin (0.35 μm), metal-coated insulating membrane attached to a silicon substrate at one end and suspended over a shallow rectangular pit. When a voltage was applied between the p+ silicon in the bottom of the pit and the deflection electrode metallization on the membrane surface, the cantilever beam experienced an electrostatic force of attraction distributed along its length, which pulled the beam downward until the membrane projection at the membrane tip made electrical contact with the fixed electrode. It became immediately apparent that these switches possessed nearly ideal electrical characteristics which made them suitable for applications requiring, for example, extremely high off-state to on-state impedance ratios, low off-state coupling capacitance, and very low switching and sustaining power.

With the maturation of MEMS technology, the MEM switch has received increased attention for application in integrated systems [3–11]. In one approach [4, 5], the switch consists of a deformable cantilever beam that realizes a segment of a signal-carrying microstrip transmission line, as shown in Figure 3.2(a). With no actuation voltage applied, the beam is undeflected and the line is open and divided into two disconnected segments. With actuation voltage applied, the beam deflects, thus closing the gap while providing a continuous path through the transmission line. In a second approach [6], a cantilever beam is also used to make or break the path through a transmission line. In this case, however, the beam is perpendicular to the transmission line it is making or breaking, as shown in Figure 3.2(b). In a third approach [7–9], a doubly supported beam/membrane is used, as shown in Figure 3.2(c). One segment of the signal-carrying transmission line originates on the substrate, right under the beam, while the other is part of the beam. Upon deflection of the beam, the gap underneath is closed, thus enabling continuity of the two line segments. In a fourth approach [10], the switch consists of a side-driven (laterally-deflecting) cantilever beam, a fixed curved driving electrode, a fixed signal electrode, and a mercury ball, deposited on the signal electrode by selective condensation, as shown in Figure 3.2(d). A voltage applied between the cantilever beam and the driving electrode causes the cantilever to deflect and close the gap between them. The curved shape of the driving electrode makes it possible to achieve a large deflection. Bumpers prevent the beam from shorting with the driving electrode. When the beam deflects enough and its tip contacts the mercury ball on the signal electrode, the switch is closed. In the last approach, [see Figure 3.2(e)], a doubly-anchored beam is disposed orthogonal to the ground (G), signal (S), ground (G) traces of a coplanar waveguide line. With no voltage applied between the beam and the bottom electrode (which coincides with the signal line), the beam is undeflected and the incoming signal passes the bridge with minimum attenuation; this is the passing, or "ON", state. On the other hand, when a voltage of the appropriate magnitude is applied between the beam and

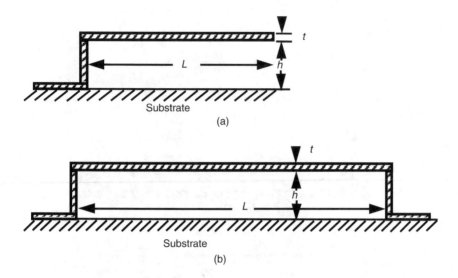

Figure 3.1 Schematic of typical surface micromachined beams: (a) cantilever beam-type, and (b) doubly supported beam. (*After:* [3].)

the bottom electrode, the beam collapses, and due to the large capacitance now loading the signal line, the incoming signal is shunted to the ground traces and reflected. Thus a highly attenuated signal ends up passing the bridge; this is the blocking, or "OFF" state.

The microwave, material, and mechanical considerations in the design of cantilever beam–type MEM switches of the type that realize a series transmission line segment [see Figure 3.2(a)], will be examined next.

3.3 MEM Switch Design Considerations

3.3.1 Switch Specifications

Ideally, switches are components that turn RF power on or off [12], or perform high-frequency signal routing [13]. It is desirable for a switch to be *noninvasive* with respect to circuit/system performance. The degree of noninvasiveness exhibited by a switch is given by its electrical parameters [13, 14].

1. *Transition time:* This is the time required for the RF voltage envelope to go from 10% to 90% for on-time, or 90% to 10% for off-time, as shown in Figure 3.3(a). At the 90% point, the signal is within 1 dB of its final value.

2. *Switching speed:* This is the time required for the switch to respond at the output when the control line input voltage changes. Switching

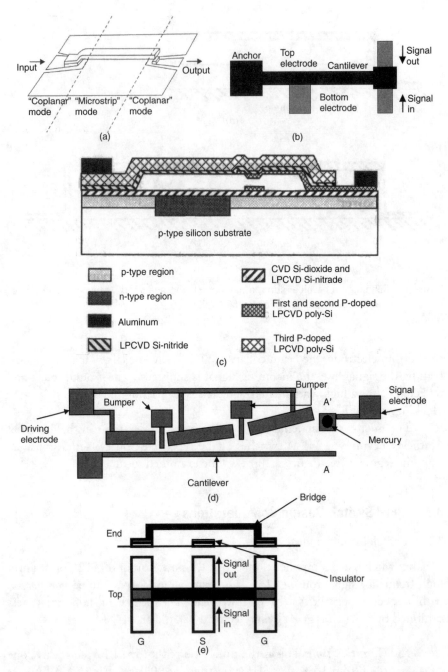

Figure 3.2 (a) Transmission line cantilever beam switch. (*From:* [4] © 1991 IEEE. Reprinted with permission.) (b) Orthogonal to signal line cantilever beam switch. (*From:* [11] © 1997 IEEE. Reprinted with permission.) (c) Doubly supported cantilever beam switch. (*From:* [7] © 1994 IEEE. Reprinted with permission.) (d) Side-driven cantilever beam switch. (*After:* [10].) (e) Shunt MEM switch.

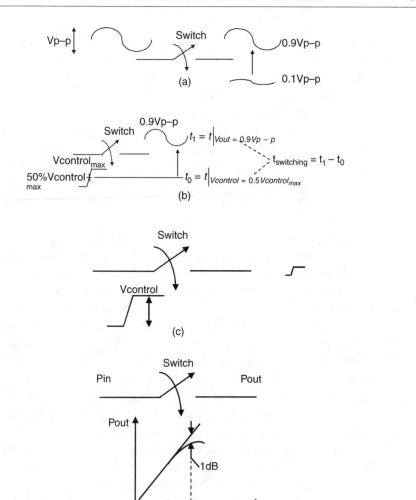

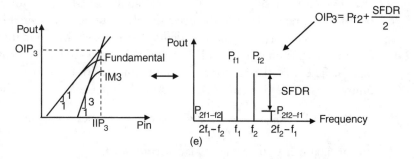

Figure 3.3 Pictorial representation of switch specifications: (a) transition time, (b) switching speed, (c) feedthrough, (d) 1-dB compression point, and (e) third-order intercept point.

speed includes the driver propagation delay as well as transition time and is measured from the 50% point on the control voltage to 90% (for the on-time) or 10% (for the off-time) of the RF voltage envelope, as shown in Figure 3.3(b). Therefore, by definition, switching time will always be longer than transition time.

3. *Switching transients:* Also called video leakage or video feedthrough, these are exponentially decaying voltage spikes at the input and/or output of a RF switch that result when the control voltage changes, as shown in Figure 3.3(c). When observed on an oscilloscope, transients are measured as the peak deviation from a steady state baseline reference.

4. *RF power handling:* This is a measure of how much and, in some respects, how well a switch passes the RF signal. To quantify RF power handling, the 1-dB compression point is commonly specified. The 1-dB compression point is a measure of the deviation from linearity of the 1-dB output power with respect to the input power, as shown in Figure 3.3(d). Alternatively, in pulsed-power operation conditions, the peak pulsed power, the repetition rate, and the duty cycle are specified. In switches containing PN-junctions (e.g., PIN diodes and MMIC switches), power handling is a function of frequency.

5. *Intercept point:* If the ratio of a switch's output power to input power (its *gain*) is a function of the input power level, then the switch is said to behave as a nonlinear device. When signals of different frequencies are simultaneously passed through the switch, then in addition to the input frequencies, the switch's output will also contain frequencies related to the sum and difference of the harmonics of the input frequencies. For example, if the input signal contains frequency components f_1 and f_2, then the output signal will *include* frequency components f_1, f_2, $2 f_1 - f_2$, and $2 f_2 - f_1$, where the last two components are called *third-order intermodulation* (IM3) *products*. The extrapolation of the distortion power to the power level of the drive signals, assuming the switch has no compression of the signals, is defined as input and output third-order intercept point, denoted IIP3 and OIP3, respectively, [see Figure 3.3(e)]. The typical unit for intercept point specification in decibel units of power relative to 1 mW, namely, the dBm:
$P(\mathrm{dBm}) = 10 \cdot \log\left(\dfrac{P(\mathrm{W})}{10^{-3}\,\mathrm{W}} \right)$. It gives a useful number from which distortion at any drive power level may be computed.

6. *RF insertion loss and isolation:* These are the loss a signal suffers upon traversing the switch, and the signal leakage appearing at the output when the switch is off, respectively.

7. *Operational lifetime:* This is the number of cycles the switch is able to operate with both dc bias voltage and RF power applied. A cycle is usually specified by a waveform that includes a dead time, the time period during which the switch is transitioning between OFF and ON states, or resting. The dead time is usually required to not exceed 5% to 10% of the cycle.

8. *Hold down time:* This is the maximum amount of time the switch will be required to remain in a given state.

9. *Operating environment:* This specifies the temperature range, the humidity level, and the radiation level the switch must withstand while operating and continuing to meet specs.

10. *Actuation power consumption:* This is the power that it takes to set the switch in a given state.

11. *Shelf life:* This is the storage time period during which the switch must maintain its integrity and functionality.

12. *Cold switching:* This refers to whether switching occurs in the absence of signal power.

13. *Hot switching:* This refers to whether switching occurs in the presence of signal power.

In addition to the above parameters, the actuation voltage, the voltage necessary to effect switching, is an important switch parameter. The following MEM switch design considerations will be based on the schematic drawing shown in Figure 3.4.

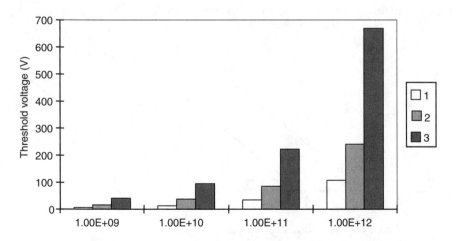

Figure 3.4 Schematic of metal-coated cantilever beam used in deflection calculations. (*From:* [11] © 1997 IEEE. Reprinted with permission.)

3.3.2 Microwave Considerations

The structure of the cantilever beam switch must be chosen so as to produce the lowest possible insertion loss, the highest possible isolation, the highest possible switching frequency, and the lowest possible actuation voltage. The insertion loss will be affected by mismatch loss, which comes from the beam's characteristic impedance being different from 50Ω, as well as losses in the beam metallization and contact resistance loss. The beam-to-substrate separation that defines the off-parasitic capacitance, on the other hand, will affect the isolation. A large beam-to-substrate separation, in turn, may result in the maximum safe deflectable distance being exceeded. Thus, due to the concomitant inverse relation between beam-to-substrate separation and material stiffness, the low off-parasitic capacitance requirement is in conflict with low actuation voltage and high switching speed (resonance frequency) requirement. The electrical design of the switch, therefore, must be carried out in a self-consistent fashion with the mechanical/material design.

For the electrical design of the MEM switch in Figure 3.2(a), we modeled it as a segment of a microstrip transmission line, as shown in Figure 3.5. Microstrip is used as interconnection media amongst devices in microwave circuits and systems where discrete devices comprising a circuit are bonded to a substrate. As the microstrip links the output of one device/circuit with the input of another, maximum power transfer considerations between circuits dictate that the microstrip's impedance, [i.e., its *characteristic* impedance (Z_0)], be set to a specific value. Through extensive electromagnetic analyses and curve fittings [15–18], a set of equations shown in (3.1) to (3.20) have been obtained, whose iterative solution gives the desired characteristic impedance as a function of conductor width, w, conductor thickness, t, substrate thickness, h, normalized conductor thickness, $t = t/(w/h)$, and substrate dielectric, ε_r. The iteration starts by

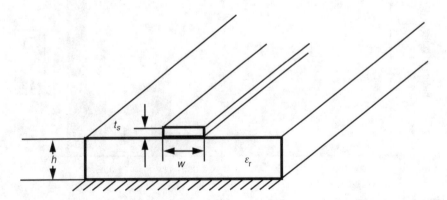

Figure 3.5 Microstrip transmission line. (*From:* [11] © 1997 IEEE. Reprinted with permission.)

obtaining an approximate value for the ratio $u = w/h$ from (3.1) and (3.2), which neglects the conductor thickness. Depending on whether the strip is embedded in a homogeneous media or in a mixed media, the ratio, u, is corrected by adding the terms Δu_a or Δu_r, respectively, as given in (3.3) to (3.6). The characteristic impedance for a microstrip in a homogeneous medium, Z_{0a}, is obtained from (3.7) and (3.8). The intrinsic wave impedance, η, is given by $\eta = \sqrt{\mu/\varepsilon_r}$, where μ and ε_r are the permeability and the permittivity of the substrate material, respectively. The corresponding characteristic impedance, Z_0, and effective dielectric constant, ε_{eff}, of the strip in the mixed medium (see Figure 3.5), are obtained from (3.9) to (3.13). Since u increases when Z_0 decreases and vice versa, one very simple and effective method for finding the new approximation for w/h is by using the ratio in (3.14). Once convergence is achieved, the conduction losses are obtained from (3.15) to (3.20). The skin-effect resistance of the conductor, given by R_m, is a function of the angular frequency, ω the permeability of the substrate, μ, and the microstrip conductivity, σ.

$$\frac{w}{h} = 8 \frac{\sqrt{\frac{A}{11}\left(7 + 4/\varepsilon_r\right) + \frac{1}{0.81}\left(1 + 1/\varepsilon_r\right)}}{A} \tag{3.1}$$

$$A = \exp\left(\frac{Z_0\sqrt{\varepsilon_r + 1}}{42.4}\right) - 1 \tag{3.2}$$

$$\Delta u_a = \frac{t}{\pi}\ln\left[1 + \frac{4\exp(1)}{t\cosh^2\sqrt{6.517w/h}}\right] \tag{3.3}$$

$$\Delta u_r = 1/2\left[1 + 1/\cosh\sqrt{\varepsilon_r - 1}\right]\Delta u_a \tag{3.4}$$

$$u_a = \frac{w}{h} + \Delta u_a \tag{3.5}$$

$$u_r = \frac{w}{h} + \Delta u_r \tag{3.6}$$

$$Z_{0a}(u) = \frac{\eta}{2\pi}\ln\left[\frac{f(u)}{u} + \sqrt{1 + \left(\frac{2}{u}\right)^2}\right] \tag{3.7}$$

$$f(u) = 6 + (2\pi - 6)\exp\left[-\left(\frac{30.666}{u}\right)^{0.7538}\right] \tag{3.8}$$

$$\varepsilon_e(u, \varepsilon_r) = \frac{\varepsilon_r + 1}{2} + \frac{\varepsilon_r - 1}{2}\left(1 + \frac{10}{u}\right)^{-s(u)b(\varepsilon r)} \tag{3.9}$$

$$a(u) = 1 + \frac{1}{49}\ln\left[\frac{u^4 + \left(\frac{u}{52}\right)^2}{u^4 + 0.432}\right] + \frac{1}{18.7}\ln\left[1 + \left(\frac{u}{18.1}\right)^3\right] \tag{3.10}$$

$$b(\varepsilon_r) = 0.564\left[\frac{\varepsilon_r - 0.9}{\varepsilon_r + 3}\right]^{0.053} \tag{3.11}$$

$$Z_0(w/t, t, \varepsilon_r) = \frac{Z_{0a}(u_r)}{\sqrt{\varepsilon_e(u_r, \varepsilon_r)}} \tag{3.12}$$

$$\varepsilon_{\text{eff}}(w/h, t, \varepsilon_r) = e_f(u_r, \varepsilon_r)\left[\frac{Z_{0a}(u_a)}{Z_{0a}(u_r)}\right]^2 \tag{3.13}$$

$$\left(\frac{w}{h}\right)_{i+1} = \left(\frac{w}{h}\right)_i \frac{Z_0(w/h, t, \varepsilon_r)}{Z_0^{desired}} \tag{3.14}$$

$$\alpha_c = \left\{0.159A\frac{R_m\left[32 - u_r^2\right]}{hZ_0\left[32 + u_r^2\right]}; w/h \le 1\right. \tag{3.15}$$

$$\alpha_c = \left\{7.02 \cdot 10^{-6} A\frac{R_m Z_0 e_{\text{eff}}}{h}\left[u_r + \frac{0667 u_r}{u_r + 1.444}\right]; \ w/h \ge 1\right. \tag{3.16}$$

$$A = 1 + u_r\left[1 + \frac{1}{\pi}\ln(2B/t)\right] \tag{3.17}$$

$$B_\ge = \left\{h; w/h \ge \frac{\pi}{2}\right. \tag{3.18}$$

$$B_{\leq} = \left\{ 2\pi w; w/h \leq \pi/2 \right. \tag{3.19}$$

$$R_m = \sqrt{\frac{\omega\mu}{2\sigma}} \tag{3.20}$$

For the electromechanical design of the MEM switch, Figure 3.4 shows a schematic of the simple, metal-coated cantilever beam, along with the parameters involved in the electrostatically actuated deflection calculations. Expressions for the threshold voltage for spontaneous beam deflection, V, the first transverse (bending) resonance frequency, f_R, and the maximum safe deflectable distance, x_{max}, are shown in (3.21) to (3.23) [19–22]. As voltage is applied, a value is reached whereby the electrostatic force at the beam tip causes the beam to become unstable, resulting in spontaneous deflection of the remaining distance until actuation. The maximum deflectable distance is the maximum distance the beam can deflect prior to failure. Failure can be designated as the maximum allowable stress, σ_{max}, that exceeds the material yield strength, σ_y. Beyond this, elastic deformation ceases, and the material fails either catastrophically (ceramic and glasses), or plastically (metals). The maximum deflectable distance must at least exceed the beam-to-substrate distance traversed by the beam.

$$V = \sqrt{\frac{2(EI)_{eff} d_{BS}{}^3 L}{\varepsilon_o \varepsilon_{r,eff} b_B l_B{}^4}} \tag{3.21}$$

$$f_R = \frac{0.162 h_B}{l_B{}^2} \sqrt{\frac{E_B K}{\rho_B}} \tag{3.22}$$

$$x_{max} = \frac{l_B{}^2 \sigma_{max}}{3 E_{eff} h'} \tag{3.23}$$

Expressions for the terms appearing in (3.21) to (3.23) are shown in (3.24) to (3.32). The $(EI)_{eff}$ term is the product of the elastic modulus, E, and the moment of inertia, I, for the equivalent beam made up of more than one material, such as the metal-coated dielectric cantilever beam. The relative contribution of the metal and the dielectric to the mechanical properties of the equivalent beam is accounted for via the expressions for K', K'', and K. The electrostatic load is integrated down the beam length, and the normalized load, L, required to produce a deflection is expressed in terms of a normalized deflection at the beam tip, Δ. To evaluate the influence of the dielectric constant on switching voltage, an $\varepsilon_{r,eff}$ term, derived from capacitors in series, is introduced as the effective dielectric constant of the dielectric beam and the void between the electrodes. However, for the case of a metal-only beam instead of a composite beam, $\varepsilon_{r,eff}$ reduces to $\varepsilon_{r,air} = 1$, and E_{eff} simplifies to E_B.

$$(EI)_{eff} = \left(\frac{b_B h_B^{\,3}}{12}\right)\frac{h_M E_B E_M}{h_M E_M + h_B E_B} K' \tag{3.24}$$

$$K' = 4 + 6\left(\frac{h_M}{h_B}\right) + \left(\frac{E_B}{E_M}\right)\left(\frac{h_B}{h_M}\right) + 4\left(\frac{h_M}{h_B}\right)^2 + \left(\frac{E_M}{E_B}\right)\left(\frac{h_M}{h_B}\right)^3 \tag{3.25}$$

$$E_{eff} = \frac{(EI)_{eff}}{I_{eff}} = \frac{12(EI)_{eff}}{b_B(h_B + h_M)^3} \tag{3.26}$$

$$L = 4\Delta^2\left[\frac{2}{3(1-\Delta)} - \frac{\tanh^{-1}\sqrt{\Delta}}{\sqrt{\Delta}} - \frac{\ln(1-\Delta)}{3\Delta}\right]^{-1} \tag{3.27}$$

$$\Delta = \frac{\delta_T}{d} \tag{3.28}$$

$$\frac{1}{\varepsilon_{r,eff}} = \frac{1}{d}\left(d_{BS} + \frac{h_B}{\varepsilon_{r,B}}\right) \tag{3.29}$$

$$K = K'K'' \tag{3.30}$$

$$K'' = \left(1 + \frac{E_B h_B}{E_M h_M}\right)\left(1 + \frac{\rho_M h_M}{\rho_B h_B}\right)^{-1} \tag{3.31}$$

$$h' = \frac{h_B + h_M}{2} \tag{3.32}$$

3.3.3 Material Considerations

The material parameters that determine the threshold voltage, the resonance frequency, and the maximum deflectable distance are the elastic modulus, the yield strength, and the dielectric constant. Table 3.1 lists these properties for different classes of materials.

The values of the mechanical and electrical properties of materials depend strongly on microstructure, such as whether the material is single crystal, poly-crystal (grain size, grain orientation), amorphous, as-deposited, or annealed [23]. These process-sensitive material properties depend on the technology and the process parameters used in the construction of the MEM switch.

Using the data shown in Table 3.1, calculations based on (3.21) and (3.32) reveal significant effects of material elastic modulus on switch perform-ance, in terms of actuation voltage and frequency. For the two sets of switch dimensions shown in Table 3.2, Figure 3.6 shows that Au, Ta_2O_5, and SiO_2 require the lowest threshold voltages, and TiO_2 and diamond require the highest threshold voltages. As E increases, the threshold voltage, V, increases. The dielectric constant does not correlate well with threshold voltage. Higher voltage is required to switch a larger elastic modulus beam material, even though the dielectric constant is higher (e.g., compare Si and SiO_2 for their ε_r, E and required V). Thus, the dielectric constant plays a minor role, but can become more influential at larger beam thicknesses.

On the other hand, for each set of switch dimensions shown in Table 3.3, Figure 3.7 shows that Ta_2O_5, Au, and SiO_2 provide the *lowest* resonance fre-quencies. Stiffer materials, however, such as Si_3N_4, TiO_2 and diamond, will exhibit higher resonance frequencies. For the most part, with all else constant, the higher the E, the higher the f_R. (A beam made of Ni has lower resonant

Table 3.1

Elastic Moduli, Yield Strengths, and Dielectric Constants of Various Materials

Class	Material	Elastic Modulus $E\,(N/m^2)$	Yield Strength $\sigma_y\,(N/m^2)$	Dielectric Constant, ε_r
Ceramics	Ta_2O_5	6.0×10^{10} (3)	—	25 (1)
	SiO_2	7.17×10^{10} (4)	8.4×10^9 (6)	3.94 (1)
	Si_3N_4	1.3×10^{11} (1)	1.4×10^{10} (4)	7 (1)
	Al_2O_3	5.3×10^{11} (4)	1.54×10^{10} (4)	10 (1)
Metals	Au	6.13×10^{10} (5)	3.24×10^8 (5)	---
	Al	7×10^{10} (4)	1.7×10^8 (4)	---
	Cr	1.8×10^{11} (1)	3.62×10^8 (4)	---
	Ni	2.07×10^{11} (4)	5.9×10^7 (4)	---
Semiconductor	Si	1.90×10^{11} (7)	7.0×10^9 (7)	13.5 (4)
Diamond	C	1.04×10^{12} (7)	5.3×10^{10} (7)	5.68 (4)

(*Source*: [11], © 1997 IEEE. Reprinted with permission.) (1) Sputtered film (various temperatures). (2) MOCVD film (500°C). (3) Electron beam evaporated film. (4) Bulk. (5) Vacuum-evaporated polycrystalline film (300°C). (6) Fiber. (7) Single crystal.

Table 3.2
Threshold Voltages for Two Sets of Beam Dimensions*

Geometrical Parameters	Sets of Dimensions	
	Set 1	Set 2
Beam length $l_B(\mu m)$	106	106
Beam width $b_B(\mu m)$	80	25
Beam height $h_B(\mu m)$	0.5	0.5
Beam-to-substrate $d_{BS}(\mu m)$	2	12

* Except for the AU beam, values are based on 50 nm-thick AU film as top electrode.

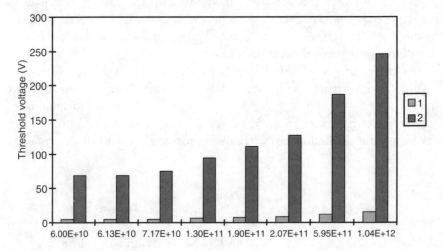

Figure 3.6 Effect of elastic modulus on threshold voltage for two sets of switch dimensions. (*From:* [11], © 1997 IEEE. Reprinted with permission.)

frequency because its density is significantly higher than the other materials with similar values of elastic modulus.)

The elastic modulus was varied over several orders of magnitude to determine its effect on voltage (see Table 3.4 and Figure 3.8). Likewise, the dielectric constant was also varied over several orders of magnitude to determine its effect on voltage (see Table 3.5 and Figure 3.9). With all else constant, decreases in E cause significant drops in V. However, increases in ε_r do not cause appreciable decrease in V, although there is a small reduction. Thus, voltage is much more sensitive to elastic modulus than to dielectric constant. The gain in reducing the voltage by selecting a higher dielectric constant material is not as great as selecting a lower elastic modulus material.

Table 3.3
Beam Dimensions Used for the Resonance Frequency Calculations Shown in Figure 3.7*

Geometrical Parameters	Sets of Dimensions		
	Set 1	Set 2	Set 3
Beam length $l_B(\mu m)$	106	80	30
Beam width $b_B(\mu m)$	25	80	30
Beam height $h_B(\mu m)$	0.5	4.0	1.2
Beam-to-substrate $d_{BS}(\mu m)$	12	1	1

(*Source*: [11], © 1997 IEEE. Reprinted with permission.)
*Except for the Au beam, values are based on 50-nm-thick Au film as top electrode.

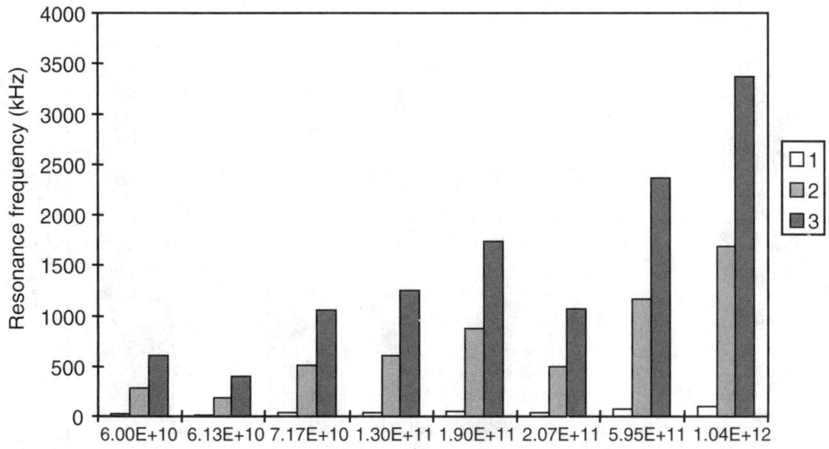

Figure 3.7 Effect of elastic modulus on resonant frequency for three sets of switch dimensions. ([11] © 1997 IEEE. Reprinted with permission.)

In addition to these material considerations, fatigue reliability of the beam poses limitations on device lifetime. For long-life fatigue design, the switch must operate at some stress level sufficiently lower than the maximum expected stress. Measures of extending the fatigue life of the material can include eliminating surface defects/discontinuities such as grain boundaries, inclusions, and pores (which can be sources for crack-initiation); and reducing the beam strain (i.e., by minimizing the beam-to-substrate distance for actuation). Amorphous microstructures are preferred over polycrystalline microstructures for long-life fatigue cycling, since grain boundaries create stress concentrations for crack-initiated fatigue failures [24].

Table 3.4
Beam Dimension Sets Used in Threshold Voltage Versus Elastic
Modulus Calculations Shown in Figure 3.8*

Geometrical Parameters	Sets of Dimensions		
	Set 1	Set 2	Set 3
Beam length $l_B(\mu m)$	106	106	106
Beam width $b_B(\mu m)$	25	25	25
Beam height $h_B(\mu m)$	2	0.5	1
Beam-to-substrate $d_{BS}(\mu m)$	2	12	12

(*Source:* [11], © 1997 IEEE. Reprinted with permission.)
*Values based on SiO_2 beam with ε_r = 3.94, and a 50 nm-thick Au film as top electrode.

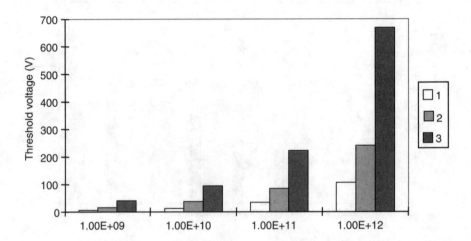

Figure 3.8 Effect of elastic modulus on threshold voltage for three sets of switch dimensions. (*From:* [11], © 1997 IEEE. Reprinted with permission.)

3.3.3.1　Contact Metallurgy

Contacts play a crucial role in the engineering of MEM switches, since their wear and tear drastically reduces switch lifetime. The key parameter characterizing contacts is their resistance, R_c, particularly in the context of a contact force, F_c [25]. Theoretical studies on contact physics [26] have established this relationship as $R_c \sim F_c^{1/3}$, when the contact force results in pure elastic deformation. Such a relationship has been verified by a number of workers [27–29], in particular, for Au, Ag, and Pd in the range of 10 μN to 1 mN. It was found that gold rendered stable electrical contacts with forces above 50　μN, and that

Table 3.5
Sample Set of Beam Dimensions Used in Threshold Voltage Versus Dielectric Constant Calculations
Shown in Figure 3.9*

Geometrical Parameters	Sets of Dimensions		
	Set 1	Set 2	Set 3
Beam length $l_B(\mu m)$	106	106	106
Beam width $b_B(\mu m)$	25	25	25
Beam height $h_B(\mu m)$	2	1	2
Beam-to-substrate $d_{BS}(\mu m)$	2	12	12

(*Source:* [11], © 1997 IEEE. Reprinted with permission.)
*Values based on SiO_2 beam with $E = 7.17 \times 10^{10} N/m^2$, and a 50-nm Au top electrode.

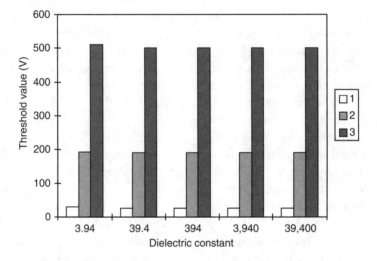

Figure 3.9 Effects of dielectric constant on voltage for three sets of switch dimensions. (*From:* [11], © 1997 IEEE. Reprinted with permission.)

contact resistances lower than 50 mΩ can be achieved by contact forces F_c lower than 1 mN for gold and gold alloys. Furthermore, it was established that pure gold contacts show the lowest resistance but tend to stick at contact forces greather than 2 mN, which happens to correspond to the release spring forces of micromachined spring (beams). Finally, it was also determined that much lower adhesion forces, between 0.1 and 0.3 mN, are exhibited by contacts with higher hardness materials like AuNi$_5$ and Rh, and that electroplated AuNi5 or hard gold (e.g. AuCo) are advantageous for microrelays.

3.3.4 Mechanical Considerations

In addition to material considerations, device performance also depends on the optimization of switch dimensions. This was examined earlier in our discussion of mechanical vibrations (see Section 2.2 of Chapter 2), which centered around the relationship between resonance frequency and geometrical/structural device parameters, (2.27), (2.37), and (2.40). In the context of its *application* as a switch, one can distinguish two performance aspects of interest, which are intimately related to the physical structure, namely, the maximum switching rate, and the switching speed. Assuming no stiction and nonhysteretic behavior, the maximum switching *rate* is in fact ideally equal to the resonance frequency for low amplitude deflections around equilibrium. The switching speed, on the other hand, is typically related to a large amplitude deflection, for example, when the excitation causes the cantilever's deflection to reach the point that triggers spontaneous deflection. In this case, the forcing function (applied voltage) is large and sudden, much like a step-function signal, and the question is how fast the beam will follow the excitation pulses. An approximate answer to this question is found in the well known result for the settling time in the transient response of a second-order system with damping ratio $\zeta = D/2\sqrt{K \cdot M}$ [30], where D is the damping coefficient, K is the spring constant and M is the mass. For a settling time corresponding to the time it takes for the beam to settle to within $\pm 5\%$ of its final configuration, this is [30]: $t_s = 3/\zeta\omega_{res} = 6M/D$. Obtaining a small settling time (high switching speed), therefore, requires reducing the structure's mass/damping ratio. This effect has been achieved by creating holes in the cantilever structure [31]. Indeed, in numerical simulations of transient responses for a cantilever with and without eight holes, Chen and Yao [31] found that the respective time to contact decreased from 8.9 ms to 67 μs, a reduction of more than 100 times.

As (3.21) and (3.22) show, a trade-off exists between voltage and resonance frequency for high device performance, such that low switching voltage also implies low switching speed. The dependencies of voltage and resonance frequency on switch dimensions are shown in (3.33) and (3.34). Both voltage and resonance frequency depend inversely on beam length, l_B. However, voltage can be reduced independently by minimizing the beam-to-substrate distance, d_{BS}. (A small beam-to-substrate distance, however, leads to high off-parasitic capacitance.) The resonance frequency can be increased independently by increasing the beam thickness h_B. Two examples of switch dimensions that lead to switching speeds in the MHz regime while requiring relatively moderate (but not low) voltages are shown in Table 3.6.

$$V \propto \sqrt{\frac{d_{BS}^3}{b_B l_B^4}} \tag{3.33}$$

Table 3.6
Sample Set of Beam Dimensions and Corresponding Calculated
Resonant Frequencies and Threshold Voltages*

Geometrical Parameters	Sets of Dimensions			
	Set 1	Set 2	Set 3	Set 4
Beam length $l_B (\mu m)$	30	30	80	80
Beam width $b_B (\mu m)$	30	30	80	80
Beam height $h_B (\mu m)$	1.2	1.2	4.0	4.0
Beam-to-substrate $d_{BS} (\mu m)$	1	1	1	1
Material	Resonance Frequencies and Threshold Voltages			
	f_{R1} (MHz)	V_1 (V)	f_{R2} (MHz)	V_2 (V)
Au	0.405	71	0.185	58
Ta_2O_5	0.594	49	0.275	28
SiO_2	1.041	59	0.513	40
Ni	1061	130	0.491	107
Si_3N_4	1.235	73	0.608	47
Si	1.723	85	0.877	52
TiO_2	2.631	143	1.167	82
Diamond	3.366	205	1.685	138

(*Source*: [11], © 1997 IEEE. Reprinted with permission.)
* Except for the Au beam, values are based on 50-nm-thick Au film as top electrode.

$$f_R \propto \frac{h_B}{l_B^2} \qquad (3.34)$$

3.3.5 Power Handling Considerations

As with every other electronics device, the maximum signal power a MEM switch can process is limited. This limitation is rooted in the following facts.

1. The conductors comprising the transmission lines and the switch contacts can only withstand a maximum current density before rupturing due to excessive heat dissipation (i.e., heating or temperature rise in the context of conductor melting point) and the onset of electromigration (i.e., the transport of mass in metals when stressed at high current densities [32]);

2. A force bias, which manifests itself as an additional electrostatic force contributing to actuation, may be elicited at high power levels by virtue of the fact that the ac voltage V applied to the system is mechanically rectified to the point of inducing a force $F = \varepsilon A V^2 / 2 d^2$ that exceeds the pull-in voltage (i.e., the signal power), rather than the controlling dc voltage, actuates the switch.

Addressing the first limitation entails determining the maximum power dissipation and electromigration parameters for the metallization system utilized, in order to avoid exceeding them. The electromigration resistance is characterized by the lifetime of the conductor lines. This lifetime is usually determined by testing a group of identical conductor lines at a certain current density and temperature, and expressing the results as a median time to failure (MTF), or t_{50}, which captures the time to reach a failure of 50%. In particular, the electromigration MTF obeys the following expression [32]:

$$t_{50} = const \cdot \frac{dw}{J^2} \exp\left(\frac{E_a}{kT} \right) \tag{3.35}$$

where w is the conductor width in cm, d is its thickness in cm, J is current density in amps/cm^2, E_a is the activation energy in eV, and T is the temperature in Kelvins. Improving resistance to electromigration is a materials-dependent affair which involves minimizing grain boundary diffusion as a result of the interaction between conductor atoms and transported current [32].

Addressing the second limitation, on the other hand, involves achieving a balance between pull-in voltage and heat dissipation in the switch structure [33]. In particular, given that the effective voltage $V_{eff} = V_{peak} / \sqrt{2}$ is related to the impedance of the system, Z_0, and the power P of the propagating signal by the equation $P = V_{peak}^2 / 2 Z_0$, it is easy to see that a power P will elicit a maximum voltage $V_{peak} = \sqrt{2 P Z_0}$. Consequently, if one desires the switch to withstand a power of, say, 10W in a 50-Ω system, then it must be designed to have a pull-in voltage $V_{Pull-in} > V_{peak} = \sqrt{2 \cdot 10W \cdot 50\Omega} = 31V$. This scenario is specifically applicable to the capacitive shunt switch, as sketched in Figure 3.10. An analysis of this structure [33, 34] determined that, in the presence of a dc voltage $V_{actuation}$, and a sinusoidal RF signal of peak amplitude V_{RF}, the effective voltage applied to the structure becomes:

$$V_{eff} = \sqrt{V_{actuation}^2 + \frac{V_{RF}^2}{2}} \tag{3.36}$$

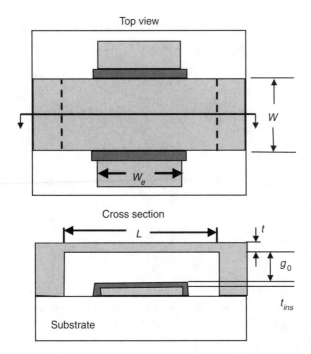

Figure 3.10 Sketch of shunt MEM switch for power-handling analysis. (*After:* [33].)

which, in terms of the signal power and system characteristic impedance, is given by:

$$V_{eff} = \sqrt{V_{actuaion}^2 + PZ_0}. \tag{3.37}$$

While the rectified voltage is developed between the bridge/ground traces and the center (signal) conductor, current does flow through the beam [34] which, due to its nonzero resistance, results in power dissipation and its concomitant temperature rise. Therefore, designing the switch structure to avoid signal-induced actuation requires engineering a pull-in voltage that exceeds V_{eff} *at the power level of interest.* In this context, the usual formula for pull-in voltage,

$$V_{Pi} = \sqrt{\frac{8k_{beam}g_0^3}{27\varepsilon_0 W W_e}} \tag{3.38}$$

must be modified [33, 34] to include power dissipation-induced heating of the bridge. The source of pull-in modification lies in the fact that the signal-induced temperature increase causes a compressive stress in the bridge, which decreases its spring constant. The temperature-dependent spring constant is given by [33]:

$$k_{beam} = \frac{32EWt^3}{L^3\left[2\left(2-\dfrac{W_e}{L}\right)\left(\dfrac{W_e}{L}\right)^2\right]} + \frac{8\sigma(1-v)Wt}{L\left(2-\dfrac{W_e}{L}\right)} \tag{3.39}$$

where E and v are the Young's modulus and Poisson's ratio for the bridge material, and σ is the stress in the beam. The temperature dependence of the stress is given by [33]:

$$\sigma = \sigma_{res} - \alpha\Delta TE\left[1 - \frac{L-W_e}{3L}\right] \tag{3.40}$$

where σ_{res} is a the residual stress resulting from the fabrication process, α is the coefficient of thermal expansion, and ΔT is the temperature rise. The temperature rise ΔT is given by the product of the power dissipation and the thermal resistance [34]:

$$\Delta T = P_{diss}R_{thermal} \tag{3.41}$$

where $R_{thermal} = R_{cond}R_{conv}/(R_{cond} + R_{conv})$, and captures the thermal resistance dependence on the assumed mechanisms of thermal dissipation, namely, conduction and convection [33]:

$$R_{cond} = \frac{L-W_e}{6kWt} \tag{3.42}$$

$$R_{cond} = \frac{1}{2hLW} \tag{3.43}$$

where k is the thermal conductivity of the bridge, and h is the convection heat transfer coefficient.

The power dissipation is given by [34]:

$$P_{diss} = \frac{1}{2}\left(2\pi V_{peak}f_{RF}C_{pi}\right)^2 R_{electrical} \tag{3.44}$$

where C_{pi} is the bridge capacitance just before it collapses, set to 1.4 C_0 ; C_0 is the capacitance of the undeflected bridge; and $R_{electrical}$ is the bridge ohmic resistance. Finally, the pull-in in the presence of power dissipation is given by [33]:

$$V_{actuation}(\Delta T) = \sqrt{V_{pi}^2(\Delta T) - Z_0 P} \tag{3.45}$$

where $V_{pi}(\Delta T)$ is obtained by substituting (3.39) and (3.40) into (3.38).

3.3.6 Reliability

While the potential of RF MEMS technology to produce switches with unique performance features is generally accepted, their adoption evidently requires improvements in reliability [14], particularly in the context of the well-established semiconductor device reliability. Fortunately, the spectacular success of the digital micromirror device (DMD) developed by Texas Instruments [35], whose lifetime has surpassed 3×10^{12} (trillion) cycles (the equivalent of over 100 years of typical office use), provides undeniable proof that acceptable MEM reliability may be attainable.

The task is indeed daunting, since a great number of factors determining reliability must be addressed. These factors, according to a report by J. Wellman [36] of the Jet Propulsion Laboratory (JPL), include:

1. Material properties, such as, fracture/failure mechanisms, elastic modulus, Poisson's ratio, fracture toughness, electrical properties (migration,etc.), interfacial strength, and coefficient of thermal expansion (CTE);

2. Issues determining residual stress, such as grain size, stiction phenomena, doping, etching parameters, surface roughness, deposition methods and parameters, postprocess release etching, and postprocess drying method (stiction);

3. Environmental effects, such as storage, humidity effects, radiation tolerance, chemical exposure effects, biocompatibility, effects of extreme heat or cold, and effects of shock.

Another issue of importance might be die singulation and packaging [37]. Naturally, assuring the reliability of MEM devices requires conducting experiments and sophisticated analyses to ascertain and gain knowledge on how these many factors affect device life. This exercise, in turn, requires the design and testing/characterization of appropriate test structures.

A number of standard test structures are usually employed for these purposes. For instance [36, 38]:

1. Doubly-anchored and cantilever beams are employed to assess residual stress, elastic modulus, pull-in voltage, coefficient of thermal expansion, and resonance frequency;

2. An array of cantilever beams of various lengths is employed to assess stiction at various steps in the process flow;

3. Four-point probe measurements are employed to assess layer resistivity, contact resistance, contact lifetime, and current capacity;

4. Capacitors are employed to assess dielectric constant.

Figure 3.11 shows stiction test structures and their profilometry analysis results.

In terms of functional device performance and reliability, automated testing and monitoring of performance degradation are crucial. Recent work by the LAAS-CNRS [39] group has evidenced an integrated setup for self-consistent

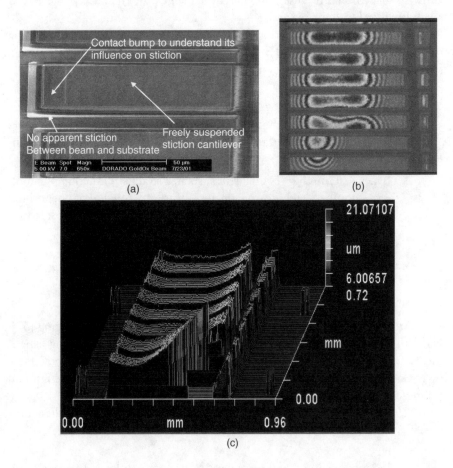

Figure 3.11 (a) Stiction test structures, (b) interferometry analysis, (c) profilometry results. (*Courtesy of:* Dr. S. Cunningham, wiSPRY, Inc.)

examination of the microwave, electrostatic, and lifetime properties of RF MEMS switches, as shown in Figure 3.12.

The setup has been employed to, among other things, monitor the effects of dielectric charging under both positive and negative swept voltage stresses under various time waveforms, as well as stiction behavior versus number of cycles.

3.4 Summary

This chapter has dealt with the design considerations for the practical utilization of the most fundamental MEM device, namely, the cantilever beam, in its application as a switch. In particular, we have addressed its operation and specifications, as well as its material, microwave, power handling, mechanical

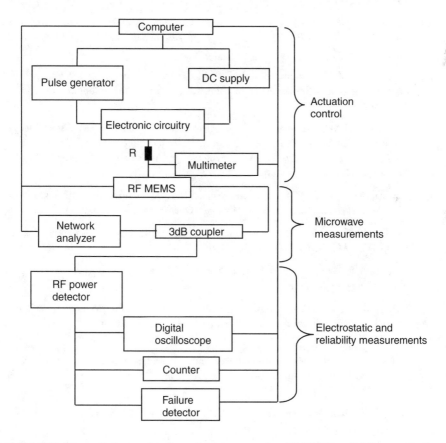

Figure 3.12 Block diagram of test equipment setup for RF MEMS reliability monitoring. (*After:* [39].)

considerations, and reliability, including the relationship between resonance frequency, switching speed, and device physical structure. The next chapter will address the other key application of the cantilever beam, namely, that of resonator for oscillator and filtering applications.

References

[1]   Petersen, K. E., "Microelectromechanical Membrane Switches on Silicon," *IBM J. Res. Develop.*, Vol. 23, No. 4, 1979, pp. 376–385.

[2]   Nathanson, H. C., et al., "The Resonant Gate Transistor," *IEEE Trans. on Electron Dev.*, Vol. 14, 1967, pp. 117–133.

[3]   Meng, Q., M. Mehregany, and R. L. Mullen, "Theoretical Modeling of Microfabricated Beams with Elastically Restrained Supports," *J. Microelectromech. Syst* , Vol. 2, No. 3, 1993, pp. 128–137.

[4]   Larson, L. E., R. H. Hackett, and R. F. Lohr, "Microactuators for GaAs-Based Microwave Integrated Circuits," *IEEE Transducers '91 Conference on Solid State Sensors and Actuators*, 1991, pp. 743–746.

[5]   Zavracky, P. M., S. Majumder, and N. E. McGruer, "Micromechanical Switches Fabricated Using Nickel Surface Micromachining," *J. Microelectromechanical Syst.*, Vol. 6, 1997, pp. 3–9.

[6]   Yao, J. J., and M. F. Chang, "A Surface Micromachined Miniature Switch for Telecommunications Applications with Signal Frequencies from dc up to 4 GHz," *Transducers '95, 8th Intl. Conf. on Solid State Sensors and Actuators, and Eurosensors IX*, Stockholm, Sweden, June 25–29, 1995, pp. 384–387.

[7]   Gretillat, M.-A., et al., "Electrostatic Polysilicon Microrelays Integrated with MOSFETs," *IEEE Microelectromechanical Systems Workshop, Oiso, Japan*, 1994, pp. 97–101.

[8]   Goldsmith, C., et al., "Micromechanical Membrane Switches for Microwave Applications," *IEEE MTT-S Digest*, 1995, pp. 91–94.

[9]   Goldsmith, C., et al., "Characteristics of Micromachined Switches at Microwave Frequencies," *IEEE MTT-S Digest*, 1996, pp. 1141–1144.

[10] Saffer, S., et al., "Mercury-Contact Switching with Gap-Closing Microantilever", *Proc. SPIE, Vol. 2882, Micromachined Devices and Components II*, Austin, TX, October 1996, pp. 204–209.

[11] de los Santos, et al., "Microwave and Mechanical Considerations in the Design of MEM Switches for Aerospace Applications," *IEEE Aerospace Conference*, Aspen, CO, February 1–8, 1997, Vol. 3, pp. 235–254.

[12] Blair, E., K. Farrington, and K. Tubbs, "Selecting the Right RF Switch," *Microwave Journal*, June 1989.

[13] Browne, J., "Switches Perform High-Frequency Signal Routing," *Microwaves & RF*, July 1989, pp. 125–132.

[14] *RF MEMS Improvement Program*, Program Research and Development Announcement, PRDA 02-07-SNK, 2002.

[15] Davis, W. A., *Microwave Semiconductor Circuit Design*, New York: Van Nostrand Reinhold Company, Inc., 1984, pp. 126–128.

[16] Wheeler, H. A., "Transmission-Line Properties of a Strip on a Dielectric Sheet on a Plane," *IEEE Microwave Theory Tech.*, Vol. 25, No. 8, 1977, pp. 631–647.

[17] Pucel, R. A., D. J. Masse, and C. P. Hartwig, "Losses in Microstrip," *IEEE Microwave Theory Tech.*, Vol. 16, No. 6, 1968, pp. 342–350; and "Corrections to 'Losses in Microstrip'," *IEEE Trans. Microwave Theory and Tech.*, Vol. 16, December 1968, p. 1064.

[18] Hammerstad, E., and O. Jensen, "Accurate Models for Microstrip Computer-Aided Design," *IEEE MTT-S Digest*, 1980, pp. 407–409.

[19] Peterson, K. E., "Dynamic Micromechanics on Silicon: Techniques and Devices," *Transactions on Electron Devices* Vol. 25, No. 10, 1978, pp. 1241–1250.

[20] Petersen, K. E., and C. R. Guarnieri, "Young's Modulus Measurements of Thin Films Using Micromechanics," *Journal of. Applied Physics*, Vol. 50, No. 11, 1979, pp. 6761–6766.

[21] Roark, R. J., and W. C. Young, *Formulas for Stress and Strain*, Fifth Ed., New York: McGraw-Hill, 1956, pp. 112–113.

[22] Peek, R. L., and H. N. Wagar, *Switching Relay Design*, Princeton, New Jersey: Van Nostrand, 1995, pp. 32–35.

[23] Dieter, G. E., *Mechanical Metallurgy*, Third Ed., New York: McGraw-Hill, 1986.

[24] Fuchs, H. O., and R. I. Stephens, *Metal Fatigue in Engineering*, New York: John Wiley and Sons, 1980.

[25] Schlaak, H. F., "Potentials and Limits of Microelectromechanical Systems forRelays and Switches," *21st International Conference on Electrical Contacts*, September 9–12, 2002, Zurich, Switzerland, pp. 19–30.

[26] Holm, R., *Electrical Contacts*, New York: Springer Verlag, 1967.

[27] Hosaka, H., H. Kuwano, and K. Yanagisawa, "Electromagnetic Microrelays: Concepts and Fundamental Characteristics," *IEEE Microelectromechanical Systems Workshop*, Fort Lauderdale, FL, February 7–10, 1993.

[28] Hyman, D., and M. Mehregany, "Contact Physics of Gold Microcontacts for MEMS Switches," *IEEE Trans. Comp. Pack. Tech.*, Vol. 22, No. 3, 1999, pp. 357–364.

[29] Kruglick, E., and K. Pister, "Lateral MEMS Microcontact Considerations," *IEEE J. Microelectromechanical Systems*, Vol. 8, No. 3, 1999, pp. 264–271.

[30] Ogata, K., *Modern Control Engineering*, Englewood Cliffs, NJ: Prentice-Hall, Inc., 1970.

[31] Chen, C.-L., and J. J. Yao, "Damping Control of MEMS Devices Using Structural Design Approach," *Solid-State Sensor and Actuator Workshop*, Hilton Head, SC, June 2–6, 1996, pp. 72–75.

[32]   Pramanik, D., and A. N. Saxena, "VLSI Metallization Using Aluminum and its Alloys," *Solid State Technology*, January 1983, pp. 127–133.

[33]   Reid, J. R., L. A. Starman, and R. T. Webster, "RF Actuation of Capacitive MEMS Switches," *2003 IEEE International Microwave Symp.*, Philadelphia, PA, June 8–16, 2003.

[34]   Rizk, J. B., E. Chaiban, and G. M. Rebeiz, "Steady State Thermal Analysis and High-Power Reliability Considerations of RF MEMS Capacitive Switches," *2003 IEEE International Microwave Symp.*, Seattle, WA, June 2–7, 2003.

[35]   Douglass, M. R., "DMD Reliability: A MEMS Success Story," *Reliability, Testing, and Characterization of MEMS/MOEMS II*, Rajeshuni Ramesham, Danelle M. Tanner, (eds.), *Proceedings of SPIE*, Vol. 4980, 2003, pp.1–11.

[36]   Wellman, J., "A General Approach to MEMS Reliability Assurance," Available on-line: http://nepp.nasa.gov/index_nasa.cfm/477/023651D8-3DE3-4F63-BC7B40DD1188831E.

[37]   Blackstone, S., "Making MEMS Reliable," *Spie's Oe Magazine*,September 2002, pp. 32–34.

[38]   Cunningham, S., "RF MEMS Fabrication Process," in *RF/Wireless MEMS Course, MEMS IV-ASME 4th Annual MEMS Technology Seminar*, Los Angeles, CA, April 26–28, 2004.

[39]   Mellé, S., et al., "Early Degradation Behavior in Parallel MEMS Switches," *4th Workshop on MEMS for MillimeterWAVE Communications (MEMSWAVE)*, Toulouse, France, July 2–4, 2003, pp. F.33–F.36.

4

Fundamental MEMS Devices: The MEM Resonator

4.1 Introduction

In this chapter we deal with the second most important application of the MEM cantilever beam, namely, that of a resonator. We have also looked at the contour-mode and wine-glass-mode disk structures, which have recently been exploited as MEM resonators. Our treatment addresses in detail its specifications, operation, both in the vertical- and lateral-displacement configurations, as well as its fundamental device design and circuit aspects in the context of its utilization in microwave oscillator and filtering applications. We will also discuss pertinent circuit models and factors limiting its ultimate performance. In the next chapter, will cover the broad application of MEMS fabrication techniques to the realization of switches and resonators, as well as to the enhancement of conventional passive microwave components, with emphasis on their fabrication and performance.

4.2 The Cantilever Beam MEM Resonator

If a time-varying excitation force is applied to a cantilever beam, its response is to vibrate. The fundamental frequency of the resulting mechanical vibration, as discussed in Chapter 2, may be described in terms of the damping coefficient, the mass, and the spring constant of the beam. Alternatively, in terms of the geometry, bulk modulus, and density of the beam, the resonance frequency of the beam is given by:

$$f_R = \frac{0.162}{l_B^2} \sqrt{\frac{E_B K}{\rho_B}} \tag{4.1}$$

The advantages of passive mechanical microresonators for applications as RF tuning elements in integrated circuits were pointed out more than 30 years ago by Nathanson, et al. [1]. In essence, mechanical resonators were found to be better than passive electronic resonators because, while the Q and resonance frequency of the former are controlled by specific material structural properties and device geometrical dimensions, those of the latter depend on difficult-to-control parameters, like IC manufacturing tolerances and temperature-induced drift.

An early demonstration of the cantilever beam mechanical microresonator was embodied in the so-called Resonant Gate Transistor (RGT) [1], as shown in Figure 4.1. The RGT consisted of three essential elements, namely: (1) an input transducer to convert the input electrical signal into a mechanical force via electrostatic attraction; (2) a mechanical microresonator; and (3) an output transducer to sense the motion of the mechanical microresonator and generate a corresponding electrical signal [e.g., a field-effect transistor (FET)]. The efforts to perfect the RGT were abandoned, however, due to two fundamental reasons, as pointed out by Lin, et al. [2]. First, the poor quality of the materials then available precluded realizing resonators with large quality factors (Qs) and small temperature-induced frequency drift; second, the nonlinear drive, characteristic of parallel plate–type transducers, was perceived as strongly constraining the input signal amplitude and dynamic range.

The subject of IC-compatible microresonators received renewed interest in the mid-to-late 1980s, when advances in the micromachining and control of polysilicon material properties became manifest [2–4]. In particular, resonant

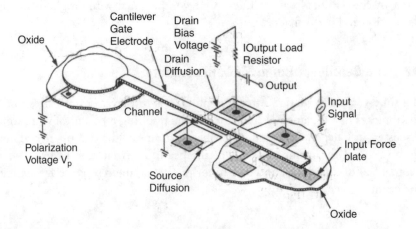

Figure 4.1 Resonant gate transistor (*From:* [1], © 1967 IEEE. Reprinted with permission.)

polysilicon beam structures for sensor applications were designed to be compatible with n-type metal oxide semiconductor (NMOS) electronic circuits technology [5–7]. The ambitious goal of this line of research is to develop IC-compatible resonators capable of displacing off-chip quartz crystal resonators.

In what follows, we address the specification and microwave considerations pertaining to these structures.

4.3 MEM Resonator Design Considerations

4.3.1 Resonator Specifications

Ideally, resonators are devices that vibrate at a specific frequency with negligible energy loss. In particular, it is desirable for a resonator to maintain its frequency of vibration despite changes in temperature, loading conditions, and age. The degree of stability exhibited by a resonator is given by its electrical parameters [8]:

1. *Center frequency* is the frequency of resonance of the first mode.

2. *Quality factor (Q)* is defined as $Q \equiv 2\pi \dfrac{Energy\ stored\ during\ a\ cycle}{Energy\ lost\ during\ the\ cycle}$.

   It is proportional to the decay time, and inversely proportional to the bandwidth around resonance. The higher the Q, the higher the frequency stability and accuracy capability of the resonator.

3. *Temperature stability, T_f,* the linear temperature coefficient of frequency, gives the temperature stability: $T_f = \dfrac{d(\log f_R)}{dT} = \dfrac{1}{f_R}\dfrac{df_R}{dT}$.

   Clearly, in light of (4.1), the temperature sensitivity in general will be a function of the expansion coefficient, the bulk modulus, the spring constant, and the material density, and can be determined for a given resonator design.

4.3.2 Microwave Considerations

4.3.2.1 Vertical Displacement Microresonator Operation

According to how they are excited, resonators are classified as either one- or two-port devices. One-port devices rely on a single electrode to both excite and detect the motion of the beam. Two-port devices, on the other hand, possess separate electrodes for exciting and detecting the mechanical vibration.

Since they embody some form of variable-gap capacitor, mechanical excitation of one-port structures, as shown in Figure 4.2, should simply entail applying a sinusoidal electrostatic drive force, which is easily accomplished by applying an ac drive voltage V_d. However, if we follow the beam displacement in time as it evolves in response to an applied *resonant* sinusoidal voltage, we make several observations. At the beginning, when the drive amplitude is zero, the displacement is zero. As the applied voltage increases to its peak positive value, the beam deflects to its maximum deflection. Then, as the driving voltage amplitude decreases towards zero, so does the displacement. Now, as the voltage continues

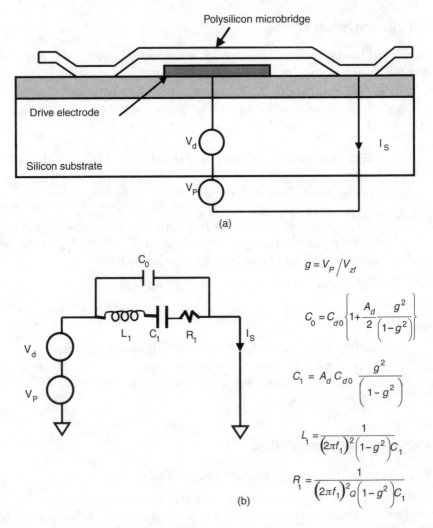

$$g = V_P / V_{zf}$$

$$C_0 = C_{d0}\left\{1 + \frac{A_d}{2}\frac{g^2}{\left(1-g^2\right)}\right\}$$

$$C_1 = A_d C_{d0}\frac{g^2}{\left(1-g^2\right)}$$

$$L_1 = \frac{1}{\left(2\pi f_1\right)^2\left(1-g^2\right)C_1}$$

$$R_1 = \frac{1}{\left(2\pi f_1\right)^2 Q\left(1-g^2\right)C_1}$$

(b)

Figure 4.2 One-port microresonator. (a) Structure, and (b) circuit model. (*From:* [7], © 1989 IEEE. Reprinted with permission.)

to decrease towards its peak negative value, the beam again *deflects towards its maximum deflection.* In other words, because the electrostatic attraction forces are at work during both the positive and negative swings of the applied voltage, the variable-gap capacitor structure reaches its maximum deflection at both peaks. Since the vibration registers two displacement peaks per cycle instead of one, this implies the structure is actually being driven at twice the intended frequency. One practical implication of this phenomenon is that the lack of an applied bias could result in the wrong oscillation frequency. To prevent this situation, it is necessary to add to the applied ac voltage a dc *polarization voltage V_p.* As long as $|V_P| > |V_d|$, only a one-polarity swing will be excited on the structure. The detection mechanism exploits the fact that, because the mechanical vibration makes the structure exhibit a time-varying capacitance, it can be detected as a displacement current flowing through the device. An approximate expression for the current through the device, given by Putty, et al. [7],

$$I_S = V_P \frac{d(\delta C_d)}{dt} + \overline{C}_d \frac{d(\delta V_d)}{dt} \tag{4.2}$$

consists of two terms. The first term is proportional to the polarization voltage, and originates from the time variation of the structure capacitance. The second term is due to the average capacitance of the device. We can immediately see from (4.2) that there is another very important motivation for adding V_P: it contributes to increased detection sensitivity by increasing the device current due to the structure motion.

4.3.2.2 One-Port Micromechanical Resonator Modeling

The analysis, design, and application of the one-port microresonator can be illustrated with a physically-based model embodying material, geometrical, and operational parameters. Such a model is shown in Figure 4.2(b) [7]. The model follows from (4.2), and consists of a series resonant circuit, which represents the device current due to the structure motion, in parallel with a capacitor, which represents device current due to the average capacitance. Five parameters specify the model, namely: (1) the mode shape factor, A_ϕ, which accounts for the distributed nature or curvature of the vibrating resonant beam; (2) the zero bias capacitance, C_{d0}, which is the device capacitance with no applied polarization voltage; (3) the resonant frequency, f_1, which is the fundamental resonant frequency of the device with no applied polarization voltage; (4) the quality factor for the driven mode, Q; and (5) the zero frequency voltage parameter, V_{zf}, which accounts for the influence of the polarization voltage on the device response.

V_{zf} is one of the most important parameters, as it characterizes the nonlinear nature of the electrostatic driving force experienced by the structure, and measures the resonator's frequency stability. Nguyen and Howe [9] quantified these effects by deriving an expression for the force exerted on the beam as function of the displacement, with due account taken for the nonlinear capacitance-displacement dependence. Specifically, if d is taken as the beam-to-electrode gap under zero bias, and x is taken as the displacement of the beam as a result of the applied bias, then the displacement-dependent capacitance is $C(x)$ given by [10]

$$C(x) = \frac{\varepsilon_0 A_{os}}{d - x} = \frac{\varepsilon_0 A_{os}}{d(1 - x/d)} = \frac{C_0}{d}\left(1 - \frac{x}{d}\right)^{-1} \tag{4.3}$$

where C_0 is the static beam-to-electrode capacitance. As the force exerted on the beam is related to the capacitance via the gradient in the stored energy [see (2.4) and (2.5)], the nonlinear nature of the capacitance induces a nonlinear force, that is,

$$U_E = \frac{1}{2}C(x) \cdot V_{applied}^2 \tag{4.4}$$

$$f_d\Big|_{\omega_0} = \frac{1}{2}\frac{\partial C(x)}{\partial x}\Big|_{\omega_0} \cdot V_{applied}^2 \tag{4.5}$$

where $\partial C/\partial x$ is a strong function of displacement [9]:

$$\frac{\partial C}{\partial x} = \frac{C_0}{d}\left(1 - \frac{x}{d}\right)^2 \tag{4.6}$$

Using (4.6), the components of force f_d at the input frequency acting on the beam of Figure 4.2 are

$$f_d\Big|_{\omega_0} = \frac{1}{2}(V_P - v_i)^2 \frac{\partial C}{\partial x}\Big|_{\omega_0} = -V_P\frac{C_0}{d}v_i + V_P^2\frac{C_0}{d^2}x \tag{4.7}$$

This equation indicates that at resonance, the force exerted on the polarized beam has two components: a first term proportional to the applied ac voltage, and a second term proportional to the instantaneous displacement. This second term, by Hooke's law, implies an electrical spring constant, $k_e = V_P^2(C_0/d^2)$, which adds to the mechanical spring constant k_m and causes the resonant frequency f_1 to be a function of the dc bias voltage V_P:

$$f_0 = \sqrt{\frac{k_m - k_e}{M}} = \sqrt{\frac{k_m}{M}\left(1 - \frac{k_e}{k_m}\right)} = f_1\sqrt{\left(1 - \frac{k_e}{k_m}\right)} \qquad (4.8)$$

Putty, et al. [7], modeled the shift in resonant frequency with polarization voltage by:

$$f_s = f_1\sqrt{1 - \left(V_P/V_{zf}\right)^2} \qquad (4.9)$$

which in a phenomenological way embodies the fact that, as the polarization voltage approaches V_{zf}, the resonant frequency of the device goes to zero. Physically, the fact that the actual value of V_{zf} surpasses the bias point for spontaneous deflection, at which the beam would no longer be vibrating [1], implies (4.9) ceases to be valid long before V_P is able to reach it. Nevertheless, (4.9) has been proven correct for its range of validity.

A plot of the typical impedance of a one-port device is shown in Figure 4.3 [7].

The device impedance is seen to be capacitive (negative phase) over most of the frequency range, and to demarcate the transitions between inductive and capacitive behavior by peaks. The lower resonant peak, corresponding to the

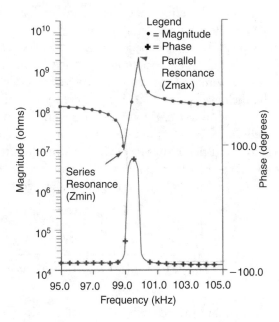

Figure 4.3 Impedance of typical one-port microresonator. (*From:* [7], © 1989 IEEE. Reprinted with permission.)

capacitive-to-inductive resonance transition, exhibits a minimum impedance, Z_{min}, of the series resonant circuit. The upper antiresonant peak, corresponding to the inductive-to-capacitive resonance transition, exhibits a maximum impedance, Z_{max}, of the parallel resonant circuit. Putty, et al. [7], summarized the relation between device impedance-frequency behavior to its physical vibration state as follows:

1. The range of inductive reactance corresponds to the resonant motion of the beam.

2. At series resonant frequency, f_s, the beam motion and the device current are at their maxima, and the device impedance is purely resistive.

3. At the parallel resonant frequency, f_p, the current due to the beam motion and the current due to the average capacitance of the device almost cancel each other, leading to a high resistive impedance.

As can be seen from an examination of the model equations included in Figure 4.2(b), since all the response features have some dependence on the polarization voltage, one would expect frequency sensitivity to variation in V_p to be a main determinant of the stability of this resonator.

4.3.2.3 Vertical Displacement Two-Port Microresonator Modeling

When the driving and detection operations are performed by separate electrodes, we obtain the two-port microresonator structure, as shown in Figure 4.4(a) [6]. Howe [5, 6] developed a linearized model, which we retrace below, to analyze and design these structures. The model assumes the structure to be excited in its fundamental mode, and involves the calculation of three quantities; namely, the drive admittance, $Y_d(j\omega)$, the forward transadmittance, $Y_f(j\omega)$, and the sense admittance, $Y_s(j\omega)$. These quantities describe the input drive voltage-to-deflection transduction, the ratio of short-circuit sense current to drive input voltage, and the output deflection-to-sense current transduction. Howe took as forward mechanical response $M_d(j\omega)$, the average deflection over the drive electrode, $\overline{U_d}(j\omega)$ in response to the electrostatic force applied by the drive electrode, $F_d(j\omega)$:

$$M_d(j\omega) = \frac{\overline{U_d}(j\omega)}{F_d(j\omega)} = \frac{a_d^2 K^{-1}}{1 - (\omega/\omega_1)^2 + j(\omega/Q\omega_1)} \tag{4.10}$$

where ω_1 is the first resonant frequency, Q is the quality factor, $K = \omega_1^2 M$ is the effective spring constant, M is the beam's mass, and a_d is the mean of the first normal mode $U_1(z)$ over the center section [6]:

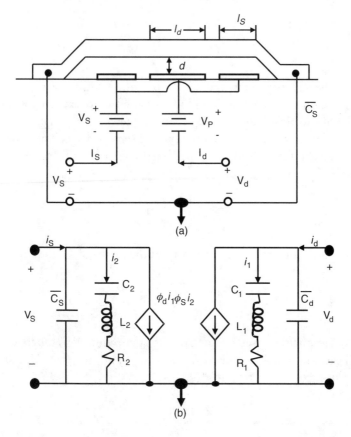

Figure 4.4 Vertical-displacement two-port resonator: (a) Structure, and (b) linearized two-port circuit model. (*After:* [6].)

$$a_d = 1.59 l_d^{-1} \left[U_1(L/2) \right]^{-1} \int_{drive} U_1(z) dz \qquad (4.11)$$

where l_d is the length of the drive electrode.

The first resonant frequency of the beam is estimated using Rayleigh's method (see Chapter 2):

$$\omega_1^2 = \left(E t^2 / 12 L^4 \right) \frac{\int_0^1 W(\zeta)\left(d^2 U_1 / d\zeta^2 \right)^2 d\zeta}{\int_0^1 \rho(\zeta) W(\zeta) U_1^2 d\zeta} \qquad (4.12)$$

where W is the beam width, L is the length, t is the thickness, ρ is the mass density, E is Young's modulus, and $\zeta = z/L$. The quality factor is given by [5]:

$$\frac{1}{Q} = \frac{\mu}{\sqrt{E\rho}} \cdot \frac{1}{t^2} \cdot \frac{1}{d^3} \cdot \left(\frac{WL}{2}\right)^2 \tag{4.13}$$

where μ is the viscosity of air, and d is the beam-to-substrate gap.

To obtain a circuit model, Howe proceeded to express the drive voltage–induced electrostatic force in terms of the drive voltage, and similarly, related the reverse mechanical response to the sense voltage. The force f_d on the beam is, as usual, found from the stored energy U_E in the drive capacitor:

$$f_d = \left(\frac{\partial U_E}{\partial \bar{u}_d}\right)_{v_d} \tag{4.14}$$

In phasor form, the component of f_d at the frequency of the drive voltage is

$$F_d(j\omega) = -\frac{\overline{C}_d}{d} V_P V_d(j\omega) + \frac{\overline{C}_d}{d^2} V_P^2 \overline{U}_d(j\omega) \tag{4.15}$$

where, for simplicity, it was assumed that the deflection $\bar{y}_d$ is much less than the nominal gap width d. The transfer function $N_d(j\omega) = \overline{U}_d(j\omega)/V_d(j\omega)$ is found by substituting (4.10) into (4.15) [6]:

$$\frac{\overline{U}_d(j\omega)}{V_d(j\omega)} = \frac{-a_d^2 K^{-1} V_P d^{-1}\left(1 - a_d^2 K^{-1}\overline{C}_d V_P^2 d^{-2}\right)}{1 - \left(\dfrac{\omega}{\omega_1\left(1 - a_d^2 K^{-1}\overline{C}_d V_P^2 d^{-2}\right)^{1/2}}\right)^2 + j\left(\dfrac{\omega}{Q\omega_1\left(1 - a_d^2 K^{-1}\overline{C}_d V_P^2 d^{-2}\right)}\right)} \tag{4.16}$$

which, defining $g_d = a_d^2 K^{-1}\overline{C}_d V_P^2 d^{-2}$, can be written as:

$$\frac{\overline{U}_d(j\omega)}{V_d(j\omega)} = \frac{-a_d^2 K^{-1} V_P d^{-1}\left(1 - g_d\right)}{1 - \left(\dfrac{\omega}{\omega_1'}\right)^2 + j\left(\dfrac{\omega}{Q'\omega_1'}\right)} \tag{4.17}$$

The resonant frequency and quality factor, which are modified by the dc polarization voltage V_p on the drive electrode, are now given by:

$$\omega_1' = \omega_1\left(1 - g_d\right)^{1/2} \tag{4.18}$$

$$Q' = Q\left(1 - g_d\right)^{1/2} \tag{4.19}$$

Now, with the deflection resulting from the drive voltage known, the drive admittance $Y_d(j\omega)$ is found by relating the drive voltage to the current into the drive capacitor $C_d(t)$ with the sense port short-circuited ($V_s(j\omega) = 0$):

$$i_d(t) = \left(\frac{\partial q}{\partial v_d}\right)_{\bar{u}_d} \frac{\partial v_d}{\partial t} + \left(\frac{\partial q}{\partial \bar{u}_d}\right)_{v_d} \frac{\partial \bar{u}_d}{\partial t} \qquad (4.20)$$

By expanding the partial derivatives and assuming the average beam displacement to be much smaller than the gap, $\bar{u}_d \ll d$, the drive current can be expressed in phasor form as follows:

$$I_d(j\omega) = j\omega \overline{C}_d V_d(j\omega) - j\omega \overline{C}_d V_p d^{-1} \overline{U}_d(j\omega) \qquad (4.21)$$

Substitution of (4.17) into (4.21) yields

$$Y_d(j\omega) = j\omega \overline{C}_d + \frac{j\omega g_d \overline{C}_d (1 - g_d)^{-1}}{1 - \left(\dfrac{\omega}{\omega_1'}\right)^2 + j\left(\dfrac{\omega}{Q'\omega_1'}\right)} \qquad (4.22)$$

To obtain the sense admittance, $Y_s(j\omega)$, a similar procedure is followed. The reverse mechanical response $M_s(j\omega)$ equals $M_d(j\omega)$, except that the parameter a_d is replaced by

$$a_s = 1.59 l_s^{-1} \left[U_1(L/2)\right]^{-1} \int_{sense} U_1(z)dz \qquad (4.23)$$

where l_d is the length of half the sense electrode, as shown in Figure 4.4(a). Howe pointed out that, because the sense electrode is located away from the center of the beam, where the maximum for excitation of the fundamental mode $U_1(z)$ occurs, $a_s < a_d$. The derivation of the output transduction factor is simplified by noting that the driving force f_s has the same form as (4.15), with the average sense capacitance $\overline{C}_S$ and the sense electrode polarization V_S replacing the $\overline{C}_d$ and V_p. Hence, the resulting expression for $\overline{Y}_s(j\omega)$ differs from (4.22) only in that $\overline{C}_S$ replaces $\overline{C}_d$ and that the parameter $g_s = a_s^2 K^{-1} \overline{C}_s V_S^2 d^{-2}$ replaces g_d.

With the above results, the forward transadmittance $Y_f(j\omega)$ is obtained as follows. The average deflection over the sense electrode $\overline{Y}_s(j\omega)$ is proportional to the sense current:

$$I_S(j\omega) = -j\omega \overline{C}_S V_S d^{-1} \overline{Y}_s(j\omega) \qquad (4.24)$$

From (4.11) and (4.23), $Y_s(j\omega)/Y_d(j\omega)$ is equal to a_s/a_d. Substitution into (4.24) of $N_d(j\omega)$ then yields

$$Y_f(j\omega) = \frac{a_s V_s \overline{C}_s}{a_d V_p \overline{C}_d} Y'_d(j\omega) = \phi_d Y'_d(j\omega) \qquad (4.25)$$

where $Y'_d(j\omega)$ is the second term in (4.22). A parallel analysis shows that the reverse transadmittance is

$$Y_r(j\omega) = \phi_s Y'_s(j\omega) \qquad (4.26)$$

where $\phi_s = \phi_d^{-1}$ and $Y'_s(j\omega)$ is defined similarly to $Y'_d(j\omega)$.

Finally, extracting the circuit elements from (4.22), (4.25), and (4.26), the linearized two-port model shown in Figure 4.4(b) is obtained. Current-controlled current sources couple the drive and sense ports by relating the motional currents i_1 and i_2 to generated currents in the other port. The final element values are

$$C_1 = g_d \overline{C}_d \left(1 - g_d\right)^{-1}$$

$$L_1 = \omega_1^{-2} \left(g_d \overline{C}_d\right)^{-1} \qquad (4.27)$$

$$R_1 = \left(g_d \omega_1 \overline{C}_d Q\right)^{-1}$$

$$C_2 = g_s \overline{C}_s \left(1 - g_s\right)^{-1}$$

$$L_2 = w_2^{-2} \left(g_s \overline{C}_s\right)^{-1} \qquad (4.28)$$

$$R_2 = \left(g_s \omega_2 \overline{C}_s Q\right)^{-1}$$

4.3.2.4   Laterally Driven Two-Port Microresonators

As mentioned previously, vertical-displacement cantilever-beam–type resonators exhibit a number of drawbacks stemming from the nonlinear dependence of the induced electrostatic force on deflection/displacement. This nonlinear dependence is, for instance, responsible for the highly undesirable polarization voltage–dependent resonance frequency. To overcome this would necessitate limiting the vibration amplitude to a small fraction of the beam-to-substrate distance, which in turn would be detrimental to the dynamic range of the system. To overcome these and other limitations, discussed later on, the laterally driven electrostatic comb-drive resonator, (see Figure 4.5), was introduced by Tang, et al. [10].

The electrostatic comb-drive resonator consists of input/output fixed electrode structures, the *stators*, and a central one-piece composite structure. The one-piece composite structure in turn consists of a folded cantilever beam and the secondary teeth, the *rotors*, which engage the input and output stators. In operation, an input ac drive voltage, together with its dc polarization voltage, is applied between the input stator, via *pad 3*, and the reference electrode *pad 1*. Notice that *pad 1* is not necessarily at ground potential. In fact, experimental studies conducted by Tang, et al. [9], indicate that resonators with compliant central folded-beam structures tend to levitate (i.e., displace normal to the substrate if they experience an imbalance in the vertically directed forces between the rotors and the stators). A natural way to attempt to minimize such an occurrence is to apply a reference voltage V_{ref} to *pad 1*, but unfortunately, doing so has the potential for causing the structure to tilt, as well as for reducing the output current. More effective approaches include reducing the ground plane area by slicing it into stripes, and configuring the comb so that drive voltages of opposite polarity are applied to adjacent electrodes over a sliced ground plane [9, 11]. The effective input drive voltage exerts an electrostatic force on the rotor. Then, since the rotor is connected to the movable folded cantilever beams, they transmit the force, and thus motion, to the beams. Consequently, the folded beams

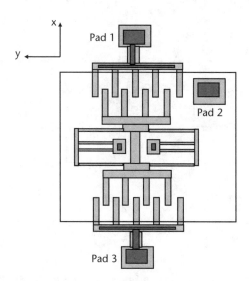

Figure 4.5 Layout of linear resonant two-port laterally-driven resonator. (*After:* [10].)

are excited into lateral vibration. Because the vibration of the input and output combs is complementary—that is, when one engages and the other disengages—the output electrode fingers sense the vibration induced on the input rotor. Furthermore, since the drive capacitance of a comb drive is linear with displacement, the force turns out to be independent of the vibration amplitude, which can be as large as $10\,\mu$m.

Extending the analysis originally performed by Howe [6], Nguyen [11] derived the circuit model for the two-port capacitively transduced resonator shown in Figure 4.6 [12]. The circuit model is linked to the material and geometrical parameters by expressions [10] for the resonance frequency, f_r,

$$f_r = \frac{1}{2\pi}\left[\frac{k_{fb}}{M_p + 0.3714M}\right]^{\frac{1}{2}} \tag{4.29}$$

the effective spring constant of the folded-beam structure, k_{fb},

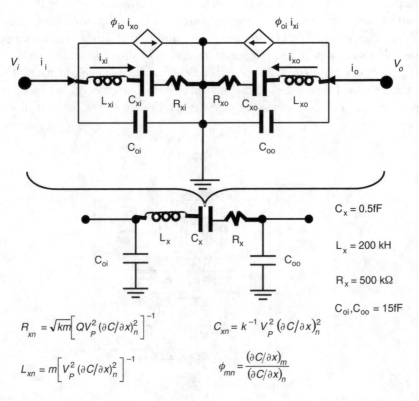

Figure 4.6 Two-port capacitively-transduced circuit model. (*From:* [12], © 1994 IEEE. Reprinted with permission.)

$$k_{fb} = \frac{24\,EI}{L^3} = 2\,Eh\left(\frac{W}{L}\right)^3 \tag{4.30}$$

where $I = (1/12)hW^3$ is the moment of inertia of the beams, and the quality factor, Q,

$$Q = \frac{d}{\mu A_p}\left(Mk_{fb}\right)^{\frac{1}{2}} \tag{4.31}$$

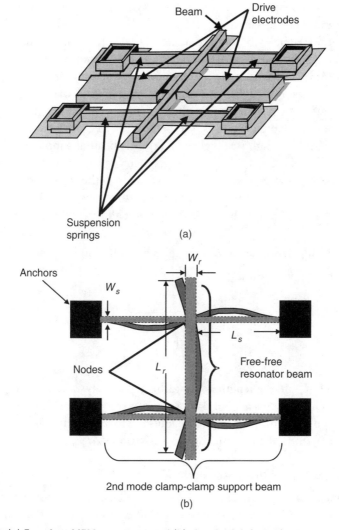

Figure 4.7 (a) Free-free MEM resonator, and (b) sketch of deformed structure under vibration. (*After:* [15].)

where μ is the absolute viscosity of the air and d is the folded-beam-to-substrate gap. As a rule, Tang, et al., pointed out that the quality factor for resonators of lateral motion tends to be much higher than that for resonators of vertical motion.

While the lateral comb-drive resonator has many desirable properties, such as large dynamic range due to the fact that the force induced by the applied signal is independent of the resulting rotor displacement [see (2.22)], its large mass precludes achieving resonance frequencies, (4.29), beyond tens of kilohertz. This limitation, together with the goal of producing megahertz-type resonance frequencies to replace quartz crystal resonators, has motivated the development of simpler/lighter microresonators. Two such resonators are the clamped-clamped beam vertical vibration resonator [13], similar to Figure 4.4, and the free-free beam lateral vibration resonator [14, 15], as shown in Figure 4.7. While the clamped-clamped resonator has demonstrated resonance frequencies of up to 8 MHz, and exhibits good dynamic range and power handling capabilities [14], it suffers from high energy dissipation to the substrates via the anchors.

This, which reflects its high stiffness, results in lower Q. To overcome the limitations of the clampled-clamped resonator, the free-free resonator was introduced [14]. In this design [see Figure 4.7(a)], the beam is supported via suspension springs [see Figure 4.7(b)], whose geometry is chosen such that when the resonator beam vibrates, the point of attachment to the suspension experiences zero displacement. This precludes the resonator acoustic energy of vibration from propagating through the suspensions through the anchors to the substrate and, thus, results in higher Q. For a resonant beam of resonance frequency $f_0 = 1.03\sqrt{EW_r / \rho L_r^2}$, where E is Young's modulus, and ρ is mass density, the length of the support beams is given by [15]

$$L_s = 1.683\left(\sqrt{\frac{EW_s}{\rho f_0}}\right)^{\frac{1}{2}} \tag{4.32}$$

where L_s and W_s are the length and width, respectively. The free-free resonator beam has demonstrated a Q of 10,743 and a resonance frequency of 10.472 MHz for a polarization voltage of 18V, and the following dimensions: $L_r = 39.8$ μm, $W_r = 2.0\,\mu$m, $L_s = 25.6\,\mu$m, and $W_s = 1.2\,\mu$m [15].

4.3.2.5   Contour-Mode Disk Resonator

In principle, attaining microresonator structures capable of much higher resonance frequencies should simply be a function of scaling down their dimensions and, thus reducing their mass. This course of action, however, leads to a number of issues having to do with increased sensitivity to the environmental disturbances (to be discussed in Section 4.3.3), but in particular, to a set of inconsistent performance parameters. A departure from this paradigm was taken when the contour-mode disk resonator was advanced, as shown in Figure 4.8. In this resonator, the vibrations occur along the disk radius. The disk is anchored at a single point at its center. One of the key advantages of this mode of vibration is that a given frequency may be achieved with larger dimensions than would be necessary with a beam. For example, a disk resonator with a radius of 17 μm, a thickness of 2 μm, a drive electrode-resonator gap of 1,000Å, and a polarization voltage of 35V, exhibited a resonance frequency of 156.23 MHz and a Q of 9,400 [16].

The design equations for the disk resonator, neglecting its thickness and finite anchor dimensions, were given by Clark, Hsu, and Nguyen [16]. In particular, the resonance frequency, f_0, is obtained by solving the following set of equations:

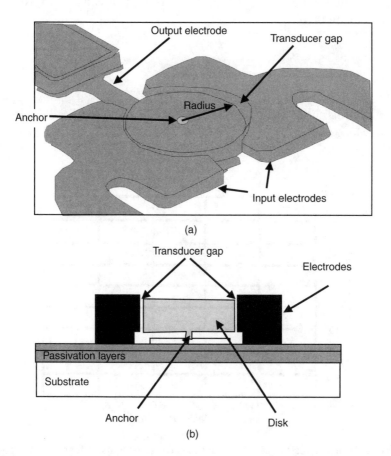

Figure 4.8 Contour-mode disk resonator. (a) Sketch. (*After:* [16].) (b) Cross-section.

$$\frac{J_0(\zeta/\xi)}{J_1(\zeta/\xi)} = 1 - \nu \tag{4.33}$$

$$\zeta = 2\pi f_0 R \sqrt{\frac{\rho(2 + 2\nu)}{E}} \tag{4.34}$$

$$\xi = \sqrt{\frac{2}{1 - \nu}} \tag{4.35}$$

where R, E, , ν and ρ are the disk radius, the the material's Young modulus, the Poisson ratio, and the mass density, respectively; $J_0(\bullet)$ and $J_1(\bullet)$ are Besel functions of the first kind or order zero and one, respectively. To provide a clearer understanding of the disk resonator, particularly with reference to the beam resonator, the solution derived from the above set of equations was approximated as follows [16]:

$$f_0 = \frac{\alpha}{R}\sqrt{\frac{E}{\rho}} \tag{4.36}$$

where α, which has a value of 0.342 for polysilicon, embodies Poisson's ratio and the mode shape.

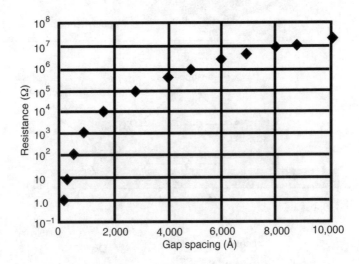

Figure 4.9 Motional resistance of contour-mode resonator as a function of resonator-electrode gap. (*After:* [16].)

Despite its virtues in terms of size, high resonance frequency, and Q, the motional resistance exhibited by contour-mode resonators is prohibitively high, as shown in Figure 4.9. This means that efficient coupling to it from a lower impedance system is impossible; at resonance, one would have to interconnect the usual system impedance of 50Ω to that of the resonators of several hundred ohms. To address this issue, development of the wine-glass–type resonator [17, 18] has been undertaken.

4.3.2.6 Wine-Glass–Mode Resonator

The wine-glass resonator has been realized as a disk [17] and as a ring [18]. The fundamental aspect underlying both, however, is the fact that a vibration mode in which the perimeter becomes oblonged displays four nodes around it, and these nodes are exploited to attach suspensions, as shown in Figure 4.10. This

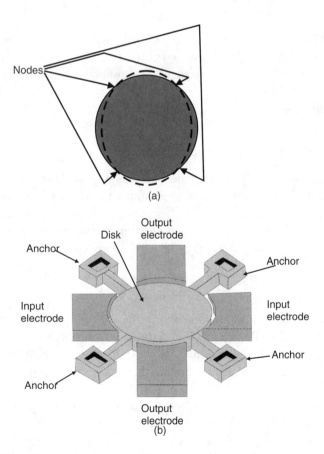

Figure 4.10 Wine-glass resonator. (a) Mode shape for disk resonator, and (b) sketch of support and driving structure. (*After:* [17].)

eliminates the need for the center anchor of the contour-mode disk, and results in a lower loss (higher Q), and, hence, lower motional resistance (4.27).

For a given vibration mode n, the resonance frequency $f_0 = \omega_0/2\pi$ of a wine-glass resonator with radius R, mass density ρ, Poisson's ratio v, and Young's modulus E, is obtained by solving the following set of mode frequency equations [17, 19]:

$$\left[\Psi_2\left(\frac{\zeta}{\xi}\right) - 2 - q\right]\left(2\Psi_2(\zeta) - 2 - q\right) = \left(nq - n\right)^2 \tag{4.37}$$

and

$$q = \frac{\zeta^2}{2n^2 - 2}, \zeta = R\sqrt{\frac{2\rho\omega_0^2(1 + v)}{E}}, \xi = \sqrt{\frac{2}{1 - v}} \tag{4.38}$$

A wine-glass resonator prototype with disk radius, R = 26.5 μm, thickness, h =1.5 μm, and disk-electrode gap, d = 1,000Å, exhibited a resonance frequency of 73.4 MHz with vacuum and atmospheric pressure Qs of 98,000 in vacuum and 8,600, respectively, together with a motional resistance of 13.23 kΩ and a polarization voltage of 7V.

4.3.3 Micromechanical Resonator Limitations

Resonators provide an essential circuit function in monolithic oscillators, filters, tuned amplifiers, and matching network applications. As these functions span the whole frequency range for RF and microwave wireless communications (e.g. from kilohertz to terahertz), a natural question to ask is whether there are any limitations that preclude their practical application. The answer, of course, is in the affirmative, and will be dealt with in this section.

4.3.3.1 Frequency Limitations

The fundamental expression for the resonance frequency of a vibrating mechanical structure of effective spring constant K, and mass M, $f_{res} \propto \sqrt{K/M}$, indicates that by reducing the mass M (i.e., by miniaturizing the device), it should be possible to obtain an arbitrarily high resonance frequency. For example, to obtain a resonance frequency of 10.7 MHz, it would be necessary to utilize a mass as small as 10^{-12} kg [12]. At these small resonator masses, however, there are two factors that come into play. First, the relative mass of the molecules populating the atmosphere in which the structure is immersed is no longer negligible. This gives rise to the phenomenon of *mass loading*, that is, the instantaneous differences in the rates of adsorption and desorption of contaminant molecules to and from the microresonator surface cause mass fluctuations, which in turn, lead to frequency fluctuations [20]. Second, the air/gas molecules that simply impinge upon the structure, without being adsorbed by it, exert a random force on it (i.e., Brownian force), which manifests itself as noise in the displacement and vibration of the structure, and thus causes fluctuations on the resonance frequency.

The factors that determine the extent to which a structure can be susceptible to mass loading have been investigated by Yong and Vig [20]. They include the adsorption and desorption rate of contaminant molecules, the contaminant molecule size and weight, the pressure, and the temperature. An expression to quantify mass loading-induced frequency fluctuations, called *phase noise density*, was advanced by Yong and Vig [20]:

$$S_\phi(f) = \frac{8 r_0 r_1 (\Delta f)^2 / N}{(r_0 + r_1)^3 + 4\pi^2 f^2 (r_0 + r_1)} \cdot \frac{1}{f^2} \qquad (4.39)$$

where r_0 is the mean rate of arrival of contaminant molecules at the microresonator site, r_1 is the desorption rate of molecules from the surface, and N is the total number of sites on the microresonator surface at which adsorption or desorption can occur.

The response of a mechanical system of mass M, spring constant K, and damping D to the Brownian noise force, has been studied by Gabrielson [21]. He found the mean squared displacement to be given by

$$\left| Z_n(f) \right|^2 = \frac{4 k_B T D}{K^2} \cdot \frac{1}{\left[1 - (f / f_{res})^2 \right]^2 + (f / f_{res})^2 / Q^2} \qquad (4.40)$$

where $4\pi^2 f_{res}^2 = \omega_{res}^2 = K/M$. The frequency fluctuations due to the Brownian noise force may be obtained by finding the spectrum of the pure sinusoidal noise-free vibration plus the noise displacement given by (4.40).

4.3.3.2 Energy Limitations

The quality factor Q of a resonator is another important factor determining its frequency stability and accuracy capability. From its definition, the Q is determined by the energy loss mechanisms in the structure, which include viscous damping for structures operating at atmospheric pressures; and intrinsic internal losses for structures operating under vacuum. Whereas structures operating at atmospheric pressure exhibit typical Qs of 100 [6], those operating under vacuum have exhibited Qs of up to 80,000 [12].

Viscous damping-related losses in vibrating structures take three forms, depending on whether the vibration displacement is normal to the substrate or parallel to it. Vibration normal to the substrate is characterized by the so-called *squeeze-film damping* effect, whereas lateral vibration losses are characterized by either *Stokes-type* or *Couette-type* damping.

Squeeze-Film Damping

Squeeze-film damping refers to the energy that must be dissipated to displace the air in the beam/plate-to-substrate gap as it vibrates vertically, as shown in Figure 4.11. This effect has been modeled either by a damping coefficient or by its corresponding damping force. For a beam with length L, much greater than its width W, and assuming an air viscosity μ, a damping coefficient given by [22]

$$D_{sf} = \frac{\mu L W^3}{d^3} \tag{4.41}$$

has been obtained. Alternatively, the equivalent damping force is given by [23]

$$F_{sf} = -\frac{\mu A W^2}{d^3} \cdot v \tag{4.42}$$

where v is the displacement velocity. The damping force can clearly be reduced by reducing the beam area, A. This, as shown in Chapter 5, takes the form of creating a grid of holes in the beam to enhance the speed of its dynamic response.

Stokes-type and Couette-type Damping

Stokes-type and Couette-type dampings come into play in the context of structures vibrating laterally, as shown in Figure 4.12. In particular, because the fluid

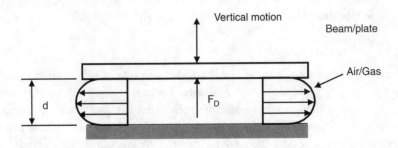

Figure 4.11 Squeeze-film damping: As the beam/plate displaces vertically, energy is expended in the process of squeezing the air film from underneath it.

ambient in which these surface-micromachined structures operate behaves as a linear slide-film damper, the viscous damping increases rapidly as the ratio of surface area to fluid thickness increases [24]. This leads to increased energy dissipation, gravely impacting the Q of the system. Cho, Pisano, and Howe [24] analyzed the viscous energy loss in oscillating fluid-film dampers that provide frictional shear for laterally driven planar microstructures. In their analysis they stipulated that Couette-type damping conditions are characterized by a linear fluid velocity profile underneath the oscillating structure, and an ambient fluid above the plate, which oscillates in time with the frequency of the plate motion, such that the viscous damping generated is negligible [24]. Assuming a damper fluid of thickness d, the estimated energy dissipated, and accompanying Q are

$$E_{diss-Couette} = \frac{\pi}{\omega} u_0^2 \left(\frac{\mu}{d}\right) \tag{4.43}$$

and

$$Q_{Couette} = \frac{\mu A}{d} \sqrt{MK} \tag{4.44}$$

where ω is the lateral oscillation angular frequency, u_0 is the peak lateral velocity, W is the strain energy stored in the system, and A is the plate area through which the energy is dissipated.

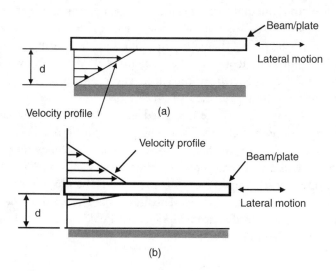

d

Velocity profile

Beam/plate

Lateral motion

(a)

Velocity profile

Beam/plate

Lateral motion

d

(b)

Figure 4.12 Energy dissipation in laterally-displaced beam/plate structures. (a) Couette-type flow, and (b) Stokes-type flow.

Cho, Pisano, and Howe's [24] analysis of Stokes-type damping, on the other hand, considered a velocity profile that embodies a strongly damped oscillation above the plate, of exponentially decaying amplitude. The ambient fluid oscillation in this case was modeled as exhibiting a phase shift proportional to the distance above the plate, and to the parameter $\beta\sqrt{\omega/2\nu}$, where ω is the circular angular frequency of plate oscillation, and ν is the kinetic viscosity. The reciprocal of β characterizes the rate of decay of the motion amplitude such that, at the distance $\delta = 1/\beta$, the amplitude has decreased by a factor e. As the energy dissipated by the Stokes-type damper, and its corresponding Q, they obtained [24]

$$E_{diss-Stokes} = \frac{\pi}{\omega} u_0^2 \mu\beta \left(\frac{\sinh 2\beta d + \sin 2\beta d}{\cosh 2\beta d - \cos 2\beta d} \right) \qquad (4.45)$$

and

$$Q_{Stokes} = \frac{Q_{Couette}}{\beta d \left(\dfrac{\sinh 2\beta d + \sin 2\beta d}{\cosh 2\beta d - \cos 2\beta d} \right)} \qquad (4.46)$$

4.3.4 Film Bulk Acoustic Wave Resonators

While research aimed at driving the resonance frequency of *silicon-compatible* mechanical microresonators to the gigahertz range continues, progress on alternative approaches, based on bulk acoustic wave cavities in piezoelectric thin films has reached the point where actual products are now being manufactured [25, 26]. The film bulk acoustic wave resonator (FBAR), as shown in Figure 4.13, is essentially a parallel-plate capacitor whose dielectric is a piezoelectric material, such as aluminum nitride (AlN) [25].

As such, when an ac voltage is applied across an FBAR, material contractions and expansions are elicited, and energy is transferred back and forth between acoustic and electric fields [26, 27]. The electrodes delimiting the acoustic cavity within which an acoustic wave bounces are spaced $\lambda/2$ apart and, in fact, embody acoustic impedance discontinuities. The physics of FBAR resonators are extensively discussed by Rosenbaum [27]. The ratio of energy stored in the electric field, U_E, to that stored in acoustic field, U_M, is called the material's *piezoelectric coupling constant*, given by:

$$\frac{U_E}{U_M} = \frac{e^2}{c^E \varepsilon^S} = K^2 \qquad (4.47)$$

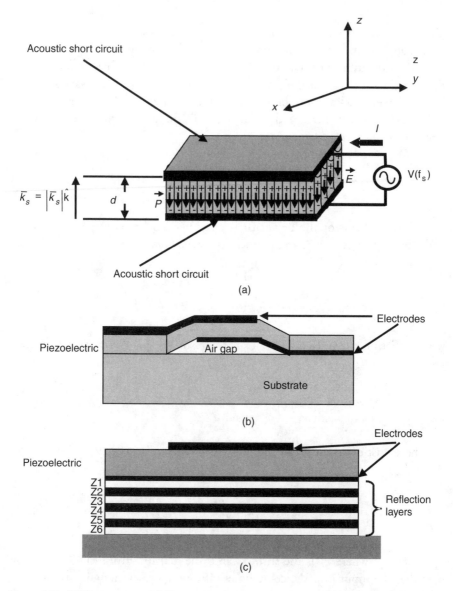

Figure 4.13 FBAR resonator. (a) Physical acoustic resonant cavity description. (*After:* [26].) (b) Membrane-supported implementation. (c) Solidly-mounted supported implementation. (*After:* [25].)

where e is the piezoelectric constant, relating material strain to induced charge flux density, c^E is the material stiffness (measured at a constant electric field), relating stress to strain (Hooke's law), and ε^S is the permittivity (measured at a constant strain). In general, these constants are tensors matrices; therefore the values to be inserted in (4.47) are associated with certain directions. These directions depend upon the normal of the electrodes across which the electric field is applied and the orientation of the piezoelectric crystal. Accordingly, two modes of excitation exist, namely, the thickness excitation (TE) mode, in which the direction of the applied electric field and the excited acoustic wave are parallel, and lateral thickness excitation (LTE) mode, in which they are perpendicular. The former is characterized by (4.47), while the latter is characterized by the *electromechanical coupling constant*, given as:

$$k_t^2 = \frac{K^2}{1 + K^2} \tag{4.48}$$

The behavior of the resonator is characterized by its impedance, which embodies both microwave and piezoelectric behaviors [27]:

$$Z_{in} = \frac{1}{j\omega C_0} \cdot \left(1 - k_t^2 \frac{\tan(kd/2)}{kd/2}\right) \tag{4.49}$$

where ω is the radian frequency, C_0 is the resonator parallel-plate capacitance, given by:

$$C_0 = \frac{\varepsilon_r \cdot \varepsilon_0 \cdot A}{d} \tag{4.50}$$

k is the acoustic wavenumber, the ratio of the radian frequency to the propagation velocity v_a:

$$k = \frac{\omega}{v_a} \tag{4.51}$$

$v_a = \sqrt{c^E/\rho}$ where c^E is the stiffness and the density. From (4.43), it can be seen that the input impedance becomes infinity, representing an antiresonance or parallel resonance whenever

$$\frac{kd}{2} = \frac{\pi}{2} \tag{4.52}$$

which occurs at a frequency ω_p, given by:

$$\omega_p = \frac{\pi v_a}{d} \cdot N \qquad N = 1, 3, 5, \ldots \tag{4.53}$$

Similarly, the series resonance occurs when the impedance is zero, when

$$1 = k_t^2 \cdot \frac{\tan(kd/2)}{kd/2} \tag{4.54}$$

Since (4.54) is a transcendental equation, in general, no closed form solution exists. However, for a "small" coupling constant, k_t, an approximate expression that relates the series resonance frequency, ω_s, to the parallel resonance frequency has been obtained, namely, [27],

$$\frac{\omega_p - \omega_s}{\omega_p} = \frac{4k_t^2}{(N\pi)^2}, \qquad N = 1,3,5,\ldots \tag{4.55}$$

The FBAR ideal circuit model is shown in Figure 4.14(a). C_0 is the parallel-plate capacitance of the resonator, and L_m and C_m determine the series and parallel resonance frequencies, respectively:

$$\omega_s = \frac{1}{\sqrt{L_m C_m}} \tag{4.56}$$

$$\omega_p = \frac{1}{\sqrt{L_m C_m}} \cdot \sqrt{1 + \frac{C_m}{C_0}} \tag{4.57}$$

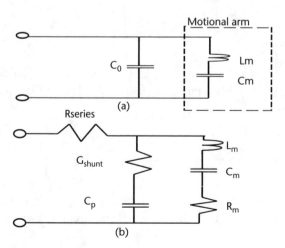

Figure 4.14 FBAR circuit model: (a) Ideal, the so-called, *Butterworth-Van Dyke Model*, and (b) realistic.

Circuit and physical models are related as follows:

$$\frac{C_m}{C_0} = \frac{8 \cdot k_t^2}{N^2 \pi^2}$$

(4.58)

and

$$L_m = \frac{\pi^3 v_a}{8\omega_s^3 \varepsilon_r \varepsilon_0 A k_t^2}$$

(4.59)

Loss effects are introduced to account for the observed finite quality factor of the resonator. In particular, since these manifest as acoustic attenuation in the metallic layers and ohmic losses, they are represented by a complex propagation constant:

$$\hat{k} = \hat{k}_r + j\hat{k}_i$$

(4.60)

where the real part is:

$$\hat{k}_r = \frac{\omega}{v_a}$$

(4.61)

and the imaginary part is:

$$\hat{k}_i = \frac{\eta\omega}{2\rho v_a^2} \cdot \left(\frac{\omega}{v_a}\right)$$

(4.62)

v_a and ρ are the acoustic viscosity and piezoelectric material density, respectively. R_m, the motional resistance, represents this loss in the equivalent circuit, and is given by:

$$R_m = \frac{\pi\eta\varepsilon_r\varepsilon_0}{8k_t^2 \rho A\omega v_a}$$

(4.63)

When losses in the leads, R_{series}, and the parallel-plate capacitor, G_{shunt}, are taken into consideration, then the more realistic circuit shown in Figure 4.14(b) results.

From the usual expression for quality factor of a series RLC circuit, namely, $Q = \omega_s L_m / R_m$, one obtains:

$$Q = \frac{v_a^2 \rho}{\omega_s \eta} \tag{4.64}$$

Relationships between the physical and circuit device models enable the inference of physical parameters from circuit parameters. The circuit parameters are obtained from a curve fitting to measured data, and a knowledge of the material density, ρ, its dielectric constant, ε_r, and its area, A, as follows:

$$k_t^2 = \frac{\pi^2}{8} \cdot \frac{C_m}{C_0} \tag{4.65}$$

$$v_a = \frac{L_m \cdot 8 \cdot \omega_s^3 \cdot \varepsilon_r \cdot \varepsilon_0 \cdot A \cdot k_t^2}{\pi^3} \tag{4.66}$$

$$c^E = v_a^2 \cdot \rho \tag{4.67}$$

$$\eta = \frac{R_m \cdot 8 \cdot k_t^2 \cdot \rho \cdot A \cdot \omega \cdot v_a}{\pi \cdot \varepsilon_r \cdot \varepsilon_0} \tag{4.68}$$

$$\eta = \frac{R_m \cdot 8 \cdot k_t^2 \cdot \rho \cdot A \cdot \omega \cdot v_a}{\pi \cdot \varepsilon_r \cdot \varepsilon_0} \tag{4.69}$$

In the context of immature, developmental processes, it is usually impossible to predict the values of process-dependent parameters such as these. Therefore, in practice the equivalent circuit information, derived from curved fitting to data from experimental devices, is employed to elucidate these fundamental device parameters.

4.4 Exercises

Exercise 4.1   *Resonator Frequency Stability*

What is the sensitivity of nominal resonator frequency, f_1, to changes in polarization voltage?

Solution:

From (4.8), we obtain the derivative of the resonance frequency with respect to the polarization voltage:

$$\frac{df_0}{dV_P} = -\frac{f_1 C_0}{dk_m} \cdot \frac{V_P}{\sqrt{1 - \frac{V_P^2 C_0}{dk_m}}}$$

Exercise 4.2 Resonator Quality Factor

Due to poor uniformity across a wafer, beams at the center of the wafer have 10% greater thickness, t, than beams near its periphery. Which beams will exhibit a higher Q? How much higher?

Solution:
From (4.13), since $Q \propto t^2$, one would expect the beams at the center to exhibit a Q approximately 20% higher than those at the periphery.

Exercise 4.3 Penetration Depth

In the Stokes-type damping model, the velocity profile exhibits strongly damped oscillations of exponentially decaying amplitude. The distance over which the velocity amplitude drops to 1% of its maximum value, u_0, is defined as the penetration depth, $\Delta = 6.48\sqrt{\nu/\omega}$. Assuming an oscillation frequency of $f = 20$ kHz, calculate Δ for a plate immersed in (a) water ($\nu = 0.01 \text{ cm}^2/s$), and (b) air ($\nu = 0.15 \text{ cm}^2/s$).

Solution:
(a) Water:

$$\Delta = 6.48 \cdot \sqrt{\frac{0.01 \frac{\text{cm}^2}{s}}{2\pi \cdot 20 \cdot 10^3 \frac{1}{s}}} = 0.001827 \text{ cm} = 18.3 \mu m$$

(b) Air:

$$\Delta = 6.48 \cdot \sqrt{\frac{0.15 \frac{\text{cm}^2}{s}}{2\pi \cdot 20 \cdot 10^3 \frac{1}{s}}} = 0.00708 \text{ cm} = 70.8 \mu m$$

Exercise 4.4 Energy Dissipation in Couette- and Stokes-type Dampers

Compare the energy dissipation effects in Couette- and Stokes-type dampers for an oscillation frequency of 80 kHz, assuming an air film thickness of (a) $d = 1$ μm, and (b) $d = 5 \mu$m.

Solution:

From (4.37) and (4.39), we obtain the ratio of Stokes- to Couette-type energy dissipation as:

$$\frac{E_{diss-Stokes}}{E_{diss-Couette}} = \beta d \left(\frac{\sinh 2\beta d + \sin 2\beta d}{\cosh 2\beta d - \cos 2\beta d} \right)$$

Now, from the frequency and viscosity data we obtain the decaying length as:

$$\beta = \sqrt{\frac{2\pi \cdot 80 \cdot 10^3 \, \frac{1}{s}}{2 \cdot 0.15 \, \frac{cm^2}{s}}} = 1294.42 \, \frac{1}{cm} = 129441.73 \, \frac{1}{m}$$

Substituting in the above equation, we obtain:
(a) For $d = 1\mu m \rightarrow \beta d = 0.129$, and we have,

$$\frac{E_{diss-Stokes}}{E_{diss-Couette}} = 0.129 \left(\frac{\sinh 0.258 + \sin 0.258}{\cosh 0.258 - \cos 0.258} \right) = 1.$$

(b) For $d = 5 \, \mu m \rightarrow \beta d = 0.647$, and we have,

$$\frac{E_{diss-Stokes}}{E_{diss-Couette}} = 0.647 \left(\frac{\sinh 1.29 + \sin 1.29}{\cosh 1.29 - \cos 1.29} \right) = 1.015.$$

4.5 Summary

This chapter has dealt with the second most important application of the MEM cantilever beam, namely, that of a resonator. We have also looked at the contour-mode and wine-glass–mode disk structures, which have recently been exploited as MEM resonators. In particular, we have addressed their operation and specifications, as well as their microwave, material, and mechanical considerations. We have also discussed pertinent circuit models and factors affecting their ultimate performance. The next chapter will cover the broad application of MEMS fabrication techniques to the realization of switches and resonators, as well to the enhancement of conventional passive microwave components, with emphasis on their fabrication and performance.

References

[1]   Nathanson, H. C., et al., "The Resonant Gate Transistor," *IEEE Trans. On Electron Dev.*, Vol. 14, 1967, pp. 117–133.

[2]   Lin, L., et al., "Microelectromechanical Filters for Signal Processing," *IEEE Microelectromechanical Systems '92*, Travemunde Germany, February 4–7, 1992, pp. 226–231.

[3]   Howe, R. T., "Polycrystalline Silicon Microstructures," *Micromachining and Micropackaging of Transducers*, C. D. Funf, et al. (eds.), New York: Elsevier Science Publishing Company Inc., 1985, pp. 169–187.

[4]   Guckel, H., et al., "Mechanical Properties of Fine Grain Polysilicon: The Repeatability Issue," Technical Digest, *IEEE Solid-State Sensor and Actuator Workshop*, Hilton Head Island, SC, June 1988, pp. 96–99.

[5]   Howe, R. T., and R. S. Muller, "Resonant Microbridge Vapor Sensor," *IEEE Trans. Electron Dev.*, Vol. 33, 1986, pp. 499–506.

[6]   Howe, R. T., "Resonant Microsensors," *Transducers '87, 4th Int. Conf. on Solid-State Sensors and Actuators, IEEE and Electrochem. Soc.*, Tokyo, Japan, 1987, pp. 843–848.

[7]   Putty, M. W., et al., "One-Port Active Polysilicon Resonant Microstructures," *IEEE Microelectromechanical Systems Workshop*, Salt Lake City, UT, 1989, pp. 60–65.

[8]   Vig, J. R., *Quartz Crystal Resonators and Oscillators for Frequency Control and Timing Applications: A Tutorial*, Army Res. Lab, Rep. No. SCLET-TR-88-1, August 1994.

[9]   Nguyen, C. T-C., and R. Howe, "Design and Performance of CMOS Micromechanical Resonator Oscillators," *IEEE Int. Freq. Control Symp.*, Boston, MA: 1994, pp. 127–134.

[10]   Nguyen, C. T-C., "Micromechanical Signal Processors" Ph.D. dissertation, U.C. Berkeley, CA, 1994.

[11]   Tang, W. C., M. G. Lim, and R. T. Howe, "Electrostatically Balanced Comb Drive for Controlled Levitation," *IEEE Solid-State Sensor and Actuator Workshop*, Salt Lake City, UT, 1989, pp. 23–27.

[12]   Tang, W. C., T.-C. H. Nguyen, and R. T. Howe, "Laterally Driven Polysilicon Resonant Microstructures," *IEEE Microelectromechanical Systems Workshop*, Salt Lake City, UT, 1989, pp. 53–59.

[13]   Bannon III, F. D., J. R. Clark, and C. T.-C. Nguyen, "High-Q HF Microelectromechanical Filters," *IEEE J. Solid-State Circuits*, Vol. 35, April 2000, pp. 512–526.

[14]   Wang. K., A.-C. Wong, and C. T-C. Nguyen, "VHF Free-Free Beam High-Q Micromechanical Resonators," *ASME/IEEE J. Microelectromechanical Systems*, Vol. 9, September 2000, pp. 347–360.

[15]   Hsu, W.-T., W.S. Best, and H. J. De Los Santos, "Design and Fabrication Procedure for High Q RF MEMS Resonators," *Microwave J.*, February 2004.

[16]   Clark, J. R., W.-T. Hsu, and C. T.-C. Nguyen, "High-Q VHF Micromechanical Contour-Mode Disk Resonators," *2000 IEEE Int. Electron Dev. Meeting*, San Francisco, Dec. 11–13, pp. 493–496.

[17] Abdelmoneum, M. A., et al., "Stemless Wine-Glass-Mode Disk Micromechanical Resonators," *2003 IEEE MEMS Conference*, Kyoto, Japan, January 19–23, pp. 698–701.

[18] Xie, Y., et al., "UHF Micromechanical Extensional Wine-Glass Mode Ring Resonators," *2003 IEEE Int. Electron Dev. Meeting.* Washington, D.C., December 4.

[19] Onoe, M., "Contour Vibrations of Isotropic Circular Plates," *J. of Acoustical Society of America*, Vol. 28, No. 6, November 1954, pp. 1158–1662.

[20] Yong, Y. K., and J. R. Vig, "Resonator Surface Contamination—A Cause of Frequency Fluctuations?" *IEEE Trans. Ultrason. Ferroelec. Freq. Contr.*, Vol. 36, No. 4, 1989, pp. 452–458.

[21] Gabrielson, T. B., "Mechanical-Thermal Noise in Micromachined Acoustic and Vibration Sensors," *IEEE Trans. Electron Dev.*, Vol. 40, No. 5, 1993, pp. 903–909.

[22] Fedder, G. K., "Simulation of Microelectromechanical Systems," Ph.D. dissertation, U.C. Berkeley, CA, 1994.

[23] Starr, J. B., "Squeeze-Film Damping in Solid-State Accelerometers," *Tech. Digest IEEE Solid State Sensor and Actuator Workshop*, Hilton Head Island, SC, June 1990, pp. 44–47.

[24] Cho, Y.-H, A. P. Pisano, and R. T. Howe, "Viscous Damping Model for Laterally Oscillating Microstructures," *IEEE J. Microelectromechanical Syst.*, Vol. 3, No. 2, 1994, pp. 81–86.

[25] Lakin, K. M., "A Review of Thin-Film Resonator Technology," *IEEE Microwave Magazine*, December 2003, pp. 61–67.

[26] McNamara, R., "FBAR Technology Shrinks CDMA Handset Duplexers," *Microwaves & RF*, September 2000, pp. 135–138.

[27] Rosenbaum, J. F., *Bulk Acoustic Wave Theory and Devices*, Norwood, MA: Artech House, 1988.

5

Microwave MEMS Applications

5.1 Introduction

The idea of employing microelectromechanical systems technology to enable complex systems containing sensing, actuating, and electronic signal processing functions in the context of a batch IC fabrication process is a rather appealing one [1–3]. Indeed, early examples of this paradigm (e.g., accelerometers for triggering air bags in automobiles, and smart pressure sensors) have become successful commercial realities [4, 5]. Among the obvious outstanding properties of these systems are their inherently smaller size and weight, and typically low power consumption [4]. As these are also highly desirable features of many microwave systems, it is natural that the subject of applying MEMS technologies to radically improve the performance of these systems is a topic that has elicited great interest. The fundamental building blocks of many microwave systems embody the following basic functions: signal sensing, generation, frequency translation, filtering, and amplification. As it is desirable to apply MEMS in a context that easily exploits the best of the micromechanical and the electronic worlds, the chosen MEM components must be compatible with established batch IC fabrication processes. Therefore, while actuation elements may be implemented using a variety of actuation mechanisms, including electromagnetic, magnetic, piezoelectric, SMA, thermoelectromechanical, and electrostatic, it is the compatibility of this latter one with surface micromachining processes [4], which singles it out as the actuation mechanism of choice.

In this chapter, we address the issue of how the unique features of MEMS fabrication technology are exploited to engineer devices with superior microwave properties. In particular, we will examine the application of MEMS in the realization of electrostatic microwave switches, transmission lines, passive lumped fixed-value, and tunable circuit elements.

5.2 MEM Switches

This section presents, in a logical fashion, a representative sample of MEMS-based switch implementations, as reflected in the literature. Our goal is to develop familiarity with specific MEM switch designs, fabrication processes, and device dimensional features, and with typical performance parameters. We present four switch topologies: the coplanar cantilever beam series switch [6], the series cantilever beam switch [7], the shunt cantilever beam switch [8], the thermal-electrostatically actuated switch [9], and the magnetic switch [10].

5.2.1 Coplanar Cantilever Beam MEM Switch Performance Analysis

In what follows, we examine both theoretical and experimental results pertaining to tradeoffs encountered in the design of MEM switches, with respect to their actuation voltage, insertion loss, isolation, and characteristic impedance.

An analysis of the performance of the coplanar cantilever beam switch, based on a Au-coated SiO_2 beam with a width of 50 μm and a length of 100-μm operating at 10 GHz, reveals that achieving a characteristic impedance close to 50Ω demands a relatively large beam-to-substrate distance, and that this condition results in a large actuation voltage (see Figures 5.1 and 5.2) [6]. On the

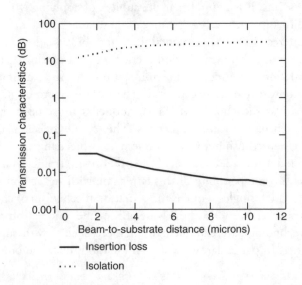

Figure 5.1 Insertion loss and isolation versus beam-to-substrate distance. Frequency is 10 GHz. (*From:* [6] © 1997 IEEE. Reprinted with permission.)

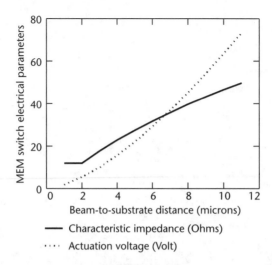

Figure 5.2 MEM switch characteristic impedance and actuation voltage versus beam-to-substrate distance. Frequency is 10 GHz. (*From:* [6], © 1997 IEEE. Reprinted with permission.)

other hand, a relatively high elastic modulus is necessary for applications requiring high frequency operation, but relatively low elastic modulus and small beam-to-substrate distance are necessary for applications requiring relatively low actuation voltages. The switch design, therefore, must consist of a self-consistent iterative design and analysis cycle, until both the electrical and mechanical requirements are met. From the point of view of the electrical characteristics of the cantilever beam switch, the beam dimensions impact the device performance greatly, as opposed to (e.g., the material properties of the beam). However, important reliability considerations, germane to aerospace applications, will be determined by the material properties, as embodied in the microstructure, the surface finish, and the residual stresses of the composite beam.

5.2.2 Electrostatic Series Cantilever Beam MEM Switch

5.2.2.1 Series Switch Operation

The series cantilever beam MEM switch has been realized in two configurations, namely, the microswitch, and the microrelay, to effect different operations, as shown in Figure 5.3. The microswitch configuration embodies a three-terminal device, similar to an FET, in which an electrostatic field applied between the beam (gate) and the bottom electrode (channel) actuates the device (i.e., causes the beam to deflect and create a path between the source and the "drain" contact).

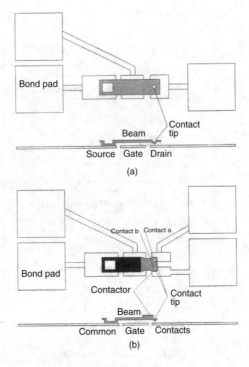

Figure 5.3 (a) Schematic drawing of a microswitch showing the source, gate, and drain. The dimple in the beam represents an indentation in the beam above the contact. (b) A microrelay showing that the actuator is separated from the contacts by an insulating layer. (*From*: [7], © 1997 IEEE. Reprinted with permission.)

In analogy with FETs, the voltage required to close the switch is called its *threshold voltage,* and the device exhibits a breakdown mechanism if the voltage between the source and the drain becomes sufficiently large. In addition, an exceedingly large drain-to-source voltage may induce an electrostatic force between the beam and drain, so as to draw the beam down and close the switch, regardless of the potential between the gate and the source. This breakdown mechanism may result in switch destruction due to excessive current, or will manifest itself as a mechanical relaxation oscillation.

5.2.2.2   Series Switch Fabrication

The process sequence for a three-terminal device, as shown in Figure 5.4, employs surface micromachining, and therefore, has four essential steps: preparation of an insulating substrate, deposition of the sacrificial layer and patterning, deposition of the structural layer, and release.

The insulating substrate used in this realization was glass or silicon coated with a 1-μm-thick SiO_2 layer. The source, gate, drain, and interconnection to

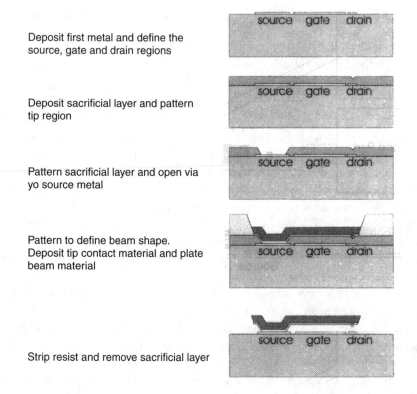

Deposit first metal and define the source, gate and drain regions

Deposit sacrificial layer and pattern tip region

Pattern sacrificial layer and open via yo source metal

Pattern to define beam shape. Deposit tip contact material and plate beam material

Strip resist and remove sacrificial layer

Figure 5.4 Process flow for the fabrication of a microswitch. (*From:* [7], © 1997 IEEE. Reprinted with permission.)

bonding pads were defined next by patterning a thin film of sputtered gold. This entailed sputtering a thin layer of chromium (200Å), to promote gold adhesion, and then 2,000Å of gold. Sacrificial material is deposited next to define the beam anchor and dimple. Two equally successful options were demonstrated as sacrificial layer material, namely, copper and aluminum. The sacrificial layer is patterned twice: first to define the beam dimple (beam tip), and second to define the contact (anchor) regions. The dimples are formed by applying an isotropic etch into the copper. The feature size expands during etching to form a semispherical indentation in the copper sacrificial layer, and may be as small as 1 μm. As long as the radius of the semispherical feature, resulting from the isotropic etch is less than the thickness of the copper layer, the indentation does not penetrate down to the substrate. The second patterning operation defines the beam anchor opening. These are vias that penetrate the copper layer, down to the Cr/Au source electrode, to facilitate gold contact/seed layer formation, and the subsequent electroplating of the beams in a nickel solution. The final step involves dissolving the sacrificial layer away, which results in releasing the beam structures.

5.2.2.3 Series Switch Performance

The switch characterization involved performing switching lifetime and actuation voltage measurements over a population of two types of devices, namely, hysteretic and nonhysteretic. No microwave characterization was pursued. To test the lifetime of nonhysteretic devices (devices designed such that no snapping is observed during beam deflection), they were subjected to switching events in room air at a rate of 2.5 kHz, while being forced to conduct load currents from 10 μA to 10 mA. To achieve these current levels, a resistor was placed in series with the device, and a power supply whose output ranged from 1 to 10V was used. Figure 5.5 shows the results for microrelays operated at an average current of 5 mA. It shows that the average survival was 3×10^6 cycles, and that all devices that experienced 10 mA survived for fewer cycles (about four orders of magnitude fewer).

The failure modes identified included gate to source shorts and contact seizure. The lifetime of the devices was observed to improve significantly when operated in a nitrogen atmosphere. Under these conditions, they were found to operate for up to 5×10^8 cycles, even when switching a 10-mA current load and tested every 1,000 cycles. In a nitrogen atmosphere, the factor limiting device lifetime was attributed to an increase in contact resistance to more than 100Ω at 2×10^9 cycles. The second measurement performed was that of threshold voltage. Nominally identical devices disposed on a single wafer with gate lengths equal to 40 μm, gate widths equal to 30 μm, and beam thicknesses of about 1.8 μm, were tested. The beam was defined by an overall beam length of 65 μm, a beam-to-substrate gap of about 1.5 μm, and a tip height of about

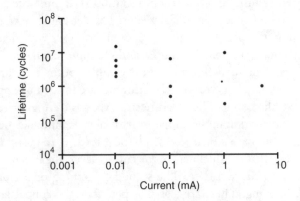

Figure 5.5 Measurements of lifetime of the microrelays as a function of the source-drain current. (*From:* [7], © 1997 IEEE. Reprinted with permission.)

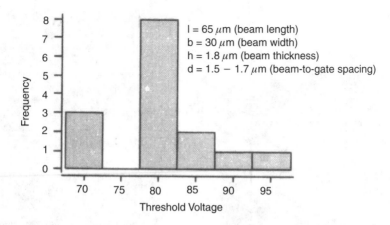

Figure 5.6 Histogram of the threshold actuation voltage for identical switches. (*From:* [7], © 1997 IEEE. Reprinted with permission.)

0.5 μm. The results are shown in the histogram of Figure 5.6. The device-to-device variation in threshold voltages is attributed to a number of factors. In the first place, there is the strain gradient in the beams, which may be responsible for a curvature of the beam tip toward the substrate of up to 0.5 μm. In the second place, there is variation in the tip height due to nonuniformities in the radii of the hemispherical dimple depth. Third, there is variation in the beam-gate spacing, due to nonuniformity in the sacrificial layer thickness. Lastly, there is variation in the beam thickness, due to variation in the structural layer thickness. Calculations of the sensitivity of the threshold voltage to the percentage deflection of the beam tip resulting from stress gradients indicate that a deflection of the tip covering 15% of the gate-to-beam gap will cause a drop in threshold voltage of close to 30%.

5.2.3 Electrostatic Shunt Cantilever Beam MEM Switch

The shunt cantilever beam MEM switch configuration, as shown in Figure 5.7, was developed to address telecommunications applications requiring a large dynamic range between ON state and OFF state impedances in the dc to 4 GHz signal frequency range, and most notably, for potential integration with MMICs. Since the lowest signal levels that can be potentially detected are set by the switch loss, it was imperative to engineer a switch with as low an insertion loss as possible. Achieving low insertion loss entails producing good electrical contacts (e.g., minimal contact resistance) when the device is ON, and a low parasitic capacitance coupling when the device is OFF. To meet these requirements, a structure consisting of a suspended silicon dioxide cantilever beam and platinum-to-gold electrical contact was adopted. In order to be compatible with

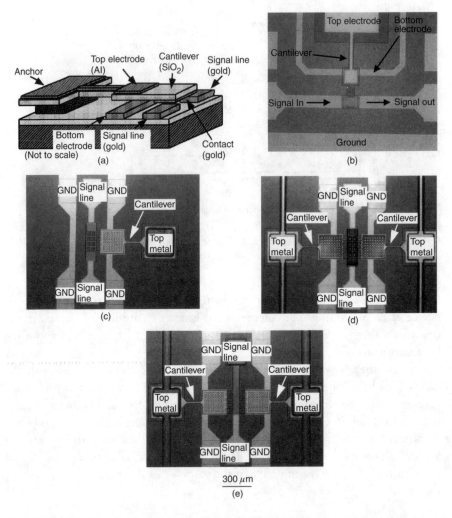

Figure 5.7 Shunt cantilever beam switch. (a) Schematic illustration, and (b) top view of micrograph image. (c), (d), and (e) Micrograph images showing the top view of three RF switch designs. (*From*: [8], © 1995 IEEE. Reprinted with permission.)

MMICs, surface micromachining on a semi-insulating GaAs substrate was adopted as the fabrication technique.

5.2.3.1 Shunt Switch Operation

In designing the switch structure, three issues must be recognized, namely, the actuation voltage, the insertion loss and isolation, and the characteristic impedance of the line. The general compromises among these parameters were discussed in Section 5.2.1. An examination of the switch structure in light of these factors reveals that, at low frequencies, the insertion loss is dominated by the ohmic losses concomitant with the signal line. This loss component is made up of two parts: the resistance of the signal line, and the contact resistance. At high frequencies, on the other hand, the insertion loss has two main causes: the resistive loss, and the skin depth effect. The skin depth effect refers to the fact that, as the conductor resistance is nonzero, the field propagating down the transmission line does not become zero exactly at the metal interface, but penetrates for a short distance into the conductor before becoming zero. As a result, this portion of the field traveling in a nonzero resistance region will incur dissipation. Quantitatively, the skin depth is defined as the distance required for the field to decay to 36.8% of its value at the interface, and is given by $\delta = 1/\sqrt{f\pi\mu\sigma}$. In this equation, f is the signal frequency, μ is the permeability of the medium surrounding the line, and σ is the conductivity of the metal making up the line. For low frequencies (e.g., less than 4 GHz), the skin depth effect may be neglected in comparison with the resistive loss of the signal line. To minimize the resistive loss, a high conductivity metal was used, namely, a thick layer (2 μm) of gold. The next structural parameter to be addressed is the width of the signal lines. The signal line width impacts three parameters, namely, the characteristic impedance, the insertion loss when the switch closed, and the isolation when it is open. Typically, as the interconnect lines in a MMIC have a characteristic impedance of 50Ω, this sets the line width. Failure to do so may introduce loss due to step width discontinuities. The width thus obtained, however, may lead to poor OFF state isolation, due to capacitive coupling across the gap of the incoming and outgoing signal lines, as shown in Figure 5.7. To reduce the capacitive coupling, and thus improve the isolation, two measures must be taken: both the gap between the gold lines and the beam-to-substrate gap must be increased. Care must be taken, however, when increasing the beam-to-substrate gap, as this will also increase the actuation voltage.

The beam design entails choosing its material, its thickness, and the thickness of the top and bottom side pad metallizations. The aluminum metallization on top of the beam forms the top of a parallel plate capacitor, and is coupled to the underlying gold bottom electrode across this gap. The actuation voltage required may, for a fixed gap distance, be tailored by adjusting the capacitor area. This measure, however, will be found not only to increase the total beam mass, but also to raise the minimum achievable switch closure time. While keeping the switch closure time constant may be achieved by attempting to increase the beam stiffness, and thus compensate for the mass increase that results from an increase in top electrode area, the required actuation voltage will be further increased. The minimum achievable insertion loss may be lowered by maximizing the thickness of the contact gold, in order to reduce the resistive loss. However, the added metal also influences the structural mass. It was based on these considerations that the final structure, to be discussed next, was arrived at.

5.2.3.2 Shunt Switch Fabrication

The fabrication technique chosen to realize the switch under discussion was surface micromachining. In this case, the substrate was a semi-insulating GaAs wafer, the sacrificial layer material was polyimide, and the structural layer was silicon dioxide, as shown in Figure 5.8.

The process starts by spinning on the sacrificial layer, which consists of three parts: a layer of thermal setting polyimide (DuPont PI2556) which defines the thickness of the bottom electrode and the signal lines, a layer of preimidized polyimide (OCG Probimide 285), and a layer of silicon nitride. Deposition of the first layer entails spinning it on, and subsequently curing it via a sequence of bakes while keeping the highest temperature of exposure below 250°C. Deposition of the second layer involves spinning it on, and baking it in a sequence of bakes while keeping the highest temperature of exposure below 170°C. Finally, a 1,500Å-thick silicon nitride layer is deposited, and patterned using photolithography and RIE in CHF_3 and O_2 chemistry.

The above is followed by opening areas on the sacrificial layer to contact the wafer surface. This is achieved via O_2 RIE, as shown in Figure 5.8(a).

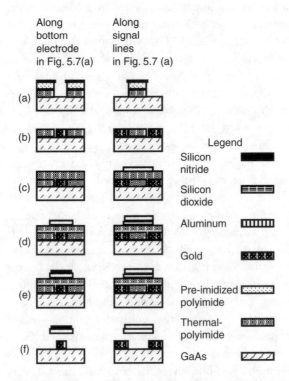

Figure 5.8 Cross-sectional schematic illustration of the shunt cantilever beam MEM switch process sequence. (*From:* [8], © 1995 IEEE. Reprinted with permission.)

To create the bottom electrode and signal line patterns, a layer of gold is deposited via electron beam evaporation, to a thickness equal to that of the thermal set Dupont polyimide layer. Pattern etching is then performed by gold lift-off, using methylene chloride to dissolve the preimidized OCG polyimide. This leaves a planar gold/polyimide surface, as shown in Figure 5.8(b). There is no danger of undercutting the cross-linked DuPont polyimide because it exhibits good chemical resistance to methylene chloride.

Next, the beam-to-substrate gap and the bottom beam contact are defined. This entails depositing a second layer of thermal-setting polyimide (DuPont PI 2555), spinning it on, and thermally cross-linking it. Then a 1-μm layer of gold is deposited by electron beam evaporation and liftoff to form the contact metal, as shown in Figure 5.8(c).

The structural beam layer is deposited next. It consists of a 2-μm thick layer of plasma-enhanced chemical vapor deposition (PECVD) silicon dioxide film. The beam is patterned using photolithography and RIE in CHF_3 and O_2 chemistry, as shown in Figure 5.8(d).

The capacitive actuator electrode plate is defined on top of the beam. To achieve this, a thin layer (2,500Å) of aluminum film is deposited by electron beam evaporation and liftoff, as shown in Figure 5.8(e). The last step is release. To dissolve the sacrificial layer, the entire switch structure is dry-etched in a Branson O_2 barrel etcher. Dry release methods are particularly recommended to preclude the potentially crippling problem of stiction, characteristic of wet chemical release approaches.

5.2.3.3 Shunt Switch Performance

The devices fabricated are defined as follows. Switches with cantilever beam lengths ranging from 100 to 200μm, and thicknesses from 10 to 2 μm, respectively, were fabricated. To assess the impact of the top electrode mass, two versions were employed, namely, a gridlike capacitor plate and a solid capacitor plate, both with an overall area of 200×200 μm. A 3-μm beam–to-signal line gap was used, while 2-μm-thick gold microstrip signal lines, with width varying from 20 to 40 μm were realized The thickness of the gold contact metal was 1 μm and the average contact area was 400μm^2.

The following switch characterization parameters were measured at atmospheric ambient: actuation voltage, switching time response, and insertion loss and isolation. The lowest measured actuation voltage was found to be 28V, with a concomitant actuation current of about 50 nA, corresponding to a switching power consumption of 1.4 μW. The time it takes the switch to close in response to an applied step in actuation voltage (i.e., the closure time), was approximately 30 μs. This measurement, in addition, unveiled the fact that, as a result of the squeeze damping effect, switches without the grid top electrode have a much larger closure and opening time. The squeeze damping effect refers to the fact that as the beam deflects and closes the beam-to-substrate gap, the air mass underneath it acts as a viscous fluid, and thus slows down the process. To assess the switching lifetime, the silicon dioxide cantilever was switched under current carrying conditions for a total of 6.5×10^{10} cycles without exhibiting any symptoms of fatigue. The maximum current handling capability for a prototype switch, with the cross-sectional dimensions of the narrowest gold signal line of $1 \mu m$ by $20 \mu m$, was 200 mA. The measured insertion loss and the isolation characteristics of the switch, measured in the 100 MHz to 4 GHz frequency range, are shown in Figure 5.9. As is clear from the figure, an insertion loss of about 0.1 dB and an isolation loss of 50 dB at 4 GHz were obtained.

5.2.4   Thermal-Electrostatic MEM Switch

While the microwave performance of electrostatic switches, such as that shown in the previous section, is excellent, the fact that they exhibit a large actuation voltage complicates their utilization in portable wireless appliances operating from a 5-V battery. To address this issue, the thermal-electrostatic MEM switch was developed [9], as shown in Figure 5.10.

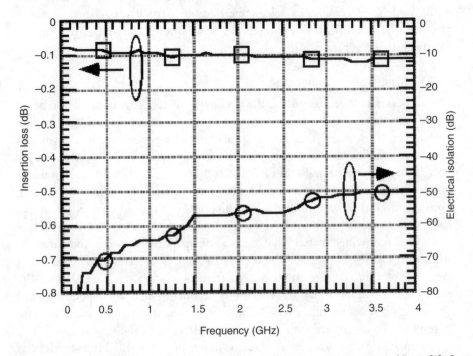

Figure 5.9 Insertion loss and isolation of shunt cantilever beam MEM switch. (*From:* [8], © 1995 IEEE. Reprinted with permission.)

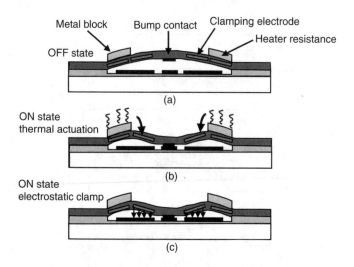

Figure 5.10 Thermal-electrostatic MEM switch: (a) Unbiased, (b) thermal actuation, (c) and electrostatic hold. (*After* [9].)

5.2.4.1 Thermal-Electrostatic Switch Operation

This switch is of the shunt type (see Chapter 3), and is therefore normally ON, as shown in Figure 5.10(a). The silicon nitride clamped–clamped beam contains titanium nitride heating resistors disposed on top next to the anchors, aluminum bimorph blocks at each end for thermal actuation, and embedded electrodes for electrostatic actuation. The switch is set into the blocking state in two steps. First, a 20-mA current, out of a 2-V supply, is passed through the resistors [see Figure 5.10(b)], which causes a temperature rise in the resistors and the aluminum/silicon nitride bimorph, resulting in downward bending of the beam, due to their different thermal expansion coefficients, until the bump contact reaches the signal line. This first step takes about 200 μs, for a total activation energy of 8 μJ. The second step of the switching process, referred to as the hold phase, involves removing the heating current and applying a dc voltage of approximately 10V between the embedded electrodes and the ground traces of the CPW lines, as shown in Figure 5.10(c). The resulting force of electrostatic attraction holds and latches the beam down. Removing the voltage causes the beam to assume its undeflected configuration within about 30 μs.

5.2.4.2 Thermal-Electrostatic Switch Fabrication

The process fabrication sequence is shown in Figure 5.11 [9]. The switch fabrication begins at the fourth planarized level of a standard silicon IC wafer.

After depositing an oxide layer via PECVD, a via is created in this layer to provide an electrical connection to the IC electronics in the wafer below, and a

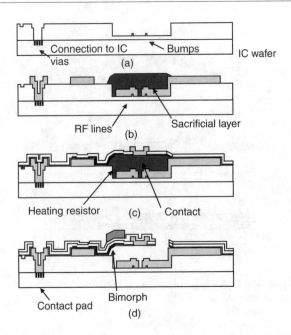

Figure 5.11 Sketch of process flow for themal-electrostatic switch fabrication. (*Source*: [9] © 2003 CEA-LETI. Courtesy of Dr. P. Robert, CEA-LETI, France.)

cavity with bumps, as shown in Figure 5.11(a). Creating these features necessitates three etching steps. Next, following seed deposition of Ni barrier and Cr adhesion layer, they deposit a 1-μm-thick gold layer, and pattern it to define contacts and RF lines. This is followed by sacrificial layer deposition to a thickness of 3 μm via spin-coating, as shown in Figure 5.11(b). Beam formation involves three steps. First, a 1,500Å low stress silicon nitride is deposited; second, a 1,500Å TiN layer is deposited for the formation of resistors and the latching electrodes; and third, a 3,000Å-thick layer of silicon nitride is deposited. This is followed by etching back 4,500Å of the nitride and a 8,000Å deposition and patterning of gold to realize the switch contact, as shown in Figure 5.11(c). Then, a 4,000Å layer of silicon nitride is deposited, followed by a 1-μm deposition and patterning of aluminum to define the bimorph actuator. Finally, access to the pads and beam contacts is provided, and the sacrificial layer is removed with an oxygen plasma, as shown in Figure 5.11(d).

5.2.4.3 Thermal-Electrostatic Switch Performance

A prototype design [9] intended for mobile communications applications and implemented in a standard silicon wafer with resistivity 15 Ω-cm exhibited, in addition to the actuation parameters of 20-mA current from a 2-V supply causing a switching time of 200 μs, a hold voltage of 15V, an insertion loss of 0.18

dB, and an isolation of 57 dB at 2 GHz. The measured lifetime was 1 billion switching cycles. Figure 5.12 shows the measured transient response of the switch.

5.2.5 Magnetic MEM Switch

None of the switches discussed thus far is bistable. Bistability, the property exhibited by a switch that possesses two stable states, is desirable when one requires that the state of the switch, ON or OFF, be maintained in the event of power loss. Ruan, Shen, and Wheeler [10] developed a bistable MEM switch by inducing changes in the magnetization of a permalloy beam.

5.2.5.1 Magnetic Switch Operation

The description and operation of the bistable magnetic MEM switch is given with reference to Figure 5.13. A beam made up of two layers, namely, a bottom metallic layer and a top magnetic layer, exhibiting a magnetic moment $\vec{m} = \vec{M}V$, where $\vec{M}$ is the magnetization and V is the beam volume, is suspended via torsional beams, [see Figure 5.13(a)]. The beam, in turn, is disposed on top of a permanent magnet, which produces a constant magnetic field H_0 perpendicular to it. This magnetic field reinforces the beam magnetization through a component $B_{0\xi} = \mu_0 H_0 \cos(\alpha)$ along the beam axis. Upon passing a current through a coil disposed on the substrate underneath the beam, the

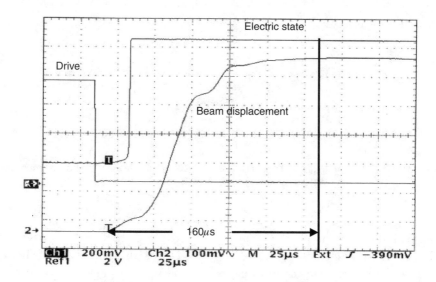

Figure 5.12 Simulation and measurement of the beam relaxation during the turn-OFF transient. (*Source*: © 2003 CEA-LETI. Courtesy of Dr. P. Robert, CEA-LETI, France.)

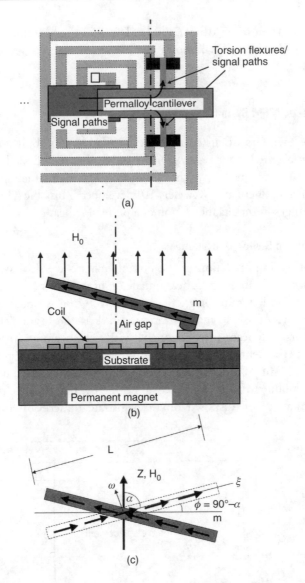

Figure 5.13 Sketch of magnetic MEM relay. (a) Top view, (b) side view, and (c) beam physics. (*After:* [10].)

magnetization direction of the beam is changed if the current-induced magnetic field along the beam axis, $B_{coil-\xi} = B_{coil-X} \sin(\alpha)$, exceeds $B_{0\xi}$, causing the beam to pivot with respect to the torsional beams and remain in that state until the coil is reenergized. Since the torsional flexures are chosen to be electrical conductors, a conduction path is established from the incoming signal line through them.

As described by Ruan, Shen, and Wheeler [10], the bistability is produced when the length, L, of a permalloy beam is much larger than its thickness, t, and width, w, in whose case the direction along its long axis L becomes that of preferred magnetization. In this situation [see Figure 5.13(c)], the beam will experience either clockwise or counterclockwise torque, depending on the initial orientation of the magnetization with respect to the external constant magnetic field H_0. In particular, if the angle α between the beam axis ξ and the external field H_0 is less than 90°, then the torque induced is counterclockwise; if the angle α between the beam axis ξ and the external field H_0 is greater than 90°, then the torque induced is clockwise. Quantitatively, the torque is given by:

$$\tau_m = \mu_0 \vec{m} \times \vec{H}_0 \tag{5.1}$$

where μ_0 is the permeability of free space, and, in order to establish a stable state, its magnitude must be greater than that of the elastic torque posed by the torsional suspensions, which is given by:

$$\tau_e = \frac{G \kappa t_s w_s^3}{L_s} \beta \tag{5.2}$$

where G is the shear modulus of the torsional spring material; κ is a constant determined by the ratio t_s / w_s; t_s, w_s, and L_s are the torsion flexure thickness, width, and length, respectively; and β is the twist angle around the flexure's section center. If the torsional flexures are located at a distance L_r from the beam's right end, then upon pivoting and switching, a contact force of magnitude

$$F \approx \frac{\tau_m}{L_r} = rwt\,\mu_0 M H_0 \sin(\alpha) \tag{5.3}$$

will be exerted, where $r = L/L_r$, and w and t are the beam's width and thickness, respectively. This force will, simultaneously, determine the contact resistance and insertion loss of the switch.

5.2.5.2 Magnetic Switch Fabrication

The magnetic switch was fabricated using conventional micromachining techniques. In particular, after first covering a silicon substrate with an insulating dielectric, the coil was made out of silver by electron-beam lithography patterning and wet etching, and covered by a polyimide layer deposited via spin-casting. Next, the bottom contact layers, made out of gold, were deposited and patterned. Using photoresist as sacrificial layer, the beam was formed by NiFe permalloy electroplating on a Au seed layer. The beam was then released and mounted on a permanent magnet.

5.2.5.3 Magnetic Switch Performance

The magnetic switch was tested functionally (i.e., no microwave performance was reported). It was pointed out by Ruan, Shen, and Wheeler [10], that a number of factors determine the performance of these devices; in particular, the rise time of the coil current, the time to realign the magnetization of the beam, the contact adhesion, and the time it takes the beam to rotate from one state to another. The prototype exhibited a driving voltage pulse of ± 5.9V with a pulse width of 0.2 ms and a corresponding current pulse of ± 79 mA for a device with beam dimensions of 800 μm $\times$ 200 μm $\times$ 25 μm, a torsion flexure located at the center, flexure dimensions 280 μm $\times$ 20 μm 3 μm, an air gap of 12 μm, and a permanent magnet with a strength of 370 Oe (0.037 T).

5.3 Micromachining-Enhanced Planar Microwave Passive Elements

The realization that large economies of scale are achievable by using batch fabrication techniques has provided a strong motivation to apply them in the implementation of planar microwave and millimeter-wave circuits. Well-known examples of technologies utilizing these fabrication techniques include MICs, which embody circuits combining planar transmission lines and packaged active and passive devices, and are typically fabricated on dielectric substrates such as Alumina (Al_2O_3, $\varepsilon_r \sim 9.8$) and Duroid^{TM} (ε_r ranging from 2.2 to 10.8); and MMICs, in which all components, both passive and active, are fabricated on semiconductor substrates such as silicon ($\varepsilon_r \sim 11.9$) or GaAs ($\varepsilon_r \sim 13.1$). In this section we introduce pertinent transmission line structures, as well as the performance issues that concern the fabrication of microwave circuits and systems on planar substrates, and which motivate tapping into micromachining techniques.

Designers have an arsenal of interconnect transmission line structures to choose from for planar applications: (1) microstrip, (2) coplanar waveguide (CPW), (3) slotline, and (4) coplanar strips. Figure 5.14 shows cross-sectional views of these lines.

Of these structures, the transmission line of choice for most designers is the microstrip, due to its ability to support an almost transverse electromagnetic (TEM) mode of propagation [11]. TEM propagation has a number of advantages. For example, it allows the neglect of high-order modes, thus rendering itself amenable to an easy approximate analysis, and to enable wideband circuits. In addition, propagation is at the velocity of light in the dielectric medium of the structure [12]. The properties of a microstrip line, as shown in Figure 5.14, can be qualitatively understood by considering its conceptual development from the familiar two-wire line, as shown in Figure 5.15 [11].

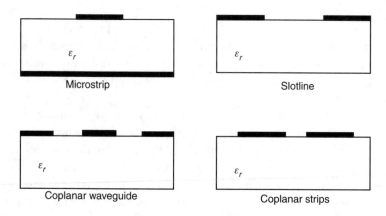

Figure 5.14 Planar transmission lines used in microwave integrated circuits. (*After:* [11].)

We begin with the transformation from (a) to (b) see Figure 5.15. This is essentially a change in the shape of the conductors. Then, if we place a conducting sheet at the plane of symmetry between the top and bottom conductors of (b), we obtain (c). Finally, if we insert a thin dielectric slab between the top and bottom conductors of (c), we obtain (d). Since the top conductor is now in an inhomogeneous environment (i.e., the dielectric constant above it is that of air while that underneath it is that of the dielectric slab), the propagation mode is no longer TEM, as in the original two-wire line, but a non-TEM hybrid mode. This is the price to be paid for planarity. In practice, the presence of non-TEM propagation means that the analysis and design of microstrip must be more

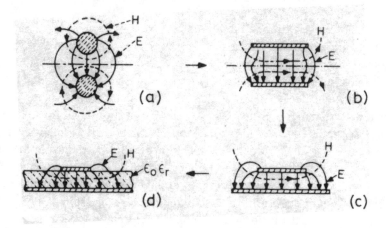

Figure 5.15 Conceptual evolution of a microstrip from two-wire line. (*After:* [11].)

rigorous, and that, because of the excitation of multiple modes, the obtainable bandwidth will be limited. The dielectric also introduces dielectric loss and frequency dispersion, which are responsible for power dissipation and unequal propagation velocity, respectively, with the latter complicating the realization of balanced circuits [11].

In addition to transmission line structures, there exist a number of passive elements like planar inductors and interdigitated capacitors, whose performance (e.g., resonance frequency), is greatly affected by the dielectric-substrate parasitic parallel-plate–type capacitance to ground. It is clear then, that these problems can be mitigated by employing the micromachining techniques developed in MEMS technology to eliminate the substrate from underneath these structures.

In this section we expose the application of each of the micromachining techniques introduced in Chapter 1, namely, surface micromachining, bulk micromachining, and LIGA to enhance the performance of transmission lines, inductors, and capacitors. For each of these items, our format proceeds as follows. We draw attention to: the physical performance limiting factors of the conventional structures; how the micromachining is applied to overcome these limitations; the actual micromachining process utilized to realize the enhanced structure; and the performance of the enhanced structure.

5.3.1 Micromachined Transmission Lines

As described in the previous section, the aim of applying micromachining technologies to remove the high dielectric substrates supporting transmission lines is to substantially reduce propagation losses, frequency dispersion, and non-TEM modes. Accordingly, we now consider a variety of transmission line structures, namely, microstrip lines, strip lines, and open and shielded coplanar lines.

5.3.1.1 Membrane-Supported Microstrip Line

The fabrication of this structure illustrates the application of the bulk micromachining technique. In this transmission line, the thin dielectric slab [13] is made up of a low dielectric constant three-layer $SiO_2/Si_3N_4/SiO_2$ supporting thin-film deposited on a high resistivity bulk silicon substrate, using thermal oxidation and high temperature chemical vapor deposition. According to Chi [14], the three-layer thin-film support is deposited as follows. The first SiO_2 layer is thermally grown at a temperature of 1,100°C to a typical thickness of 5,000Å. The middle Si_3N_4 layer is deposited by low pressure chemical vapor deposition (LPCVD) at 820°C to a typical thickness of 3,000Å, with gas flow rates of 160 sccm (cm^2/min, standard temperature and pressure) ammonia (NH_3), and 40 sccm dichlorosilane (DCS) [14]. The top SiO_2 layer is then deposited by LPCVD at a temperature of 920°C to a typical thickness 4,000Å, with gas flow rates of 290 sccm (N_2), 120 sccm(N_2O), and 60 sccm(DCS). After this step, the substrate's top and back sides are covered by $SiO_2/Si_3N_4/SiO_2$ membranes. After the membrane is deposited the transmission line stripe is defined on the top side of the substrate. This is accomplished by first depositing a seed layer of gold via evaporation, then depositing additional gold via electroplating to satisfy skin depth thickness requirements, and finally defining the stripe using conventional photolithography. Bulk micromachining is invoked at this point to open the membrane on the backside underneath the line, and then to etch the bulk of the substrate until the dielectric membrane is reached. Chi [14] recommends the use of two etchants, depending on the nature of the substrate, namely, potassium hydroxide (KOH) or ethylene diamine-pyrocatechol water (EDP) for silicon substrates, and a sulfuric acid/peroxide/water-based solution ($H_2SO_4/H_2O_2/H_2O$) or RIE for GaAs substrates. Figure 5.16 shows a bulk micromachined microstrip transmission line fabricated using this approach. Because the microstrip metal trace is supported by a microthin low-dielectric constant membrane, the line is very close to being homogeneous. This results in the propagation of nearly TEM modes, virtually negligible dielectric loss, and an extremely wide single mode bandwidth. For example, it has been reported that a membrane which is 500 μm wide on a 350-μm-substrate sustains a pure TEM mode from dc to about 320 GHz, whereas a similar line on silicon will propagate a higher order mode at 62 GHz [13]. One drawback accompanying the near-unity dielectric constant membrane, however, is that no advantage can be taken of the $\sqrt{\varepsilon_r}$ size reduction factor; thus circuits with relatively large dimensions are obtained.

5.3.1.2 Microshield Transmission Line

A drawback of the membrane-supported microstrip transmission line discussed above, is that, as it possesses no intrinsic ground plane, the structure must be placed on top of another metallized substrate. This requires some form of assembly process (e.g., soldering or fusion bonding), which makes the application of

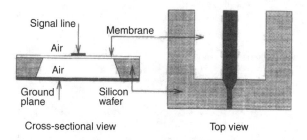

Cross-sectional view   Top view

Figure 5.16 Illustration of membrane-supported microstrip transmission line. At 30 GHz a 106Ω microstrip on membrane (14 mil) exhibits a loss of 0.025 dB/mm. (*From:* [13], © 1996 IEEE. Reprinted with permission.)

the line somewhat labor-intensive. It is desirable, therefore, to consider alternative line structures which, while availing themselves of the benefits enjoyed by the membrane-supported line, can function with a ground plane *coplanar* with the stripe. Such an alternative is the *coplanar microshield* line [15], as shown in Figure 5.17.

The coplanar microshield line is an extension of the membrane-supported line, in that the membrane is now populated by the center conductor and ground planes. There is no need to worry about setting the membrane-to-substrate distance in order to produce a given characteristic impedance value, because all impedance-determining parameters, namely, the center stripe width, S, the center stripe-to-ground-plane gap, W, and the ground plane to ground plane distance, G, can be defined photolithographically on the top surface. Addition of a metallized lower shielding cavity has been reported to minimize signal crosstalk between adjacent lines (hence the name *microshield*), and to eliminate radiation into parasitic substrate modes. Since the inner surface of the micromachined bulk is metallized, the top and bottom ground planes may be directly connected. This is advantageous, compared to microstrip and the conventional coplanar line, in that it renders via holes or air bridges to connect to ground unnecessary [15]. The performance of microshield lines has been characterized in terms of their attenuation characteristics. Figure 5.18 shows experimental and theoretical results, which compare the performance of microshield and conventional coplanar lines [15].

The measured results validated the intuition regarding microshield line performance. They showed it to be free from unexpected or excessive conductor loss mechanisms, and to be comparable in performance to the conventional coplanar waveguide at lower frequencies. At high frequencies (e.g., millimeter wavelengths), the line displayed no dielectric-related loss as well as nondispersive, single-mode propagation. Finally, the measured effective dielectric constant of two microshield structures is shown in Figure 5.19 [15].

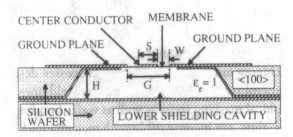

Figure 5.17 The microshield transmission line geometry (not to scale). The cross-hatched lines indicate metallization, which is typically 1 to 2 μm thick. (*From:* [15], © 1995 IEEE. Reprinted with permission.)

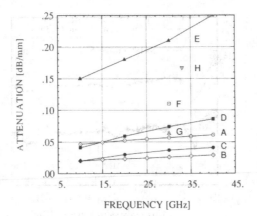

Curve	Line	ϵ_r	Substr	S	W	H	t	Z_O Ω	Data	Ref.
A	μshield	1.0	Air	250	25	355	1.2	75	Meas	–
B	μshield	1.0	Air	190	55	355	1.2	100	Meas	–
C	CPW	12.8	GaAs	232	84	100	-	50	Calc	18
D	CPW	12.8	GaAs	69	28	100	-	50	Calc	18
E	CPW	12.9	GaAs	88	16	500	1.0	30	Meas	19
F	CPW	12.9	GaAs	250	25	500	1.0	30	Calc	16
G	CPW	4.0	Quartz	250	25	250	1.0	50	Calc	16
H	GCPW	11.7	Si	50	125	355	1.2	73	Meas	–

Figure 5.18 Attenuation for microshield and coplanar waveguide lines. *S* is the center conductor width, *W* is the slot width, *H* is the substrate height, and *t* is the metal thickness (in μm). The width of the lower shielding cavity for the microshield is 1,800 μm. (*From*: [15], © 1995 IEEE. Reprinted with permission.)

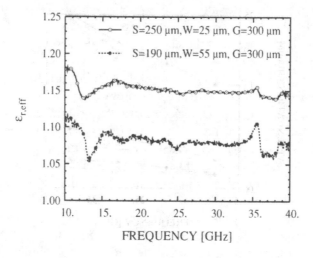

Figure 5.19 Measured effective dielectric constant, $\varepsilon_{r,eff}$ on two microshield lines with different aspect ratios. The cavity height (*H*) is 350 μm. (*From*: [15], © 1995 IEEE. Reprinted with permission.)

The measured dielectric constant is very close to unity, which indicates a high degree of success in eliminating the substrate from underneath the line. The fact that the $\varepsilon_{r,eff}$ is not exactly unity, however, does reflect in the tendency to increase slightly with frequency, as is to be expected for an inhomogeneous dielectric.

5.3.1.3 Top-Side-Etched Coplanar Waveguide

The micromachining and metal deposition processes described above for the fabrication of membrane microstrip and microshield lines require many photo-lithographical masking steps, as well as wafer bonding and backside processing; thus, they are incompatible with mainstream IC manufacturing processes. An alternative transmission line structure, whose fabrication avoids backside processing [16], is the conventional *coplanar* line. Milanovic, et al. [16], utilized top-side etching bulk micromachining to fabricate coplanar transmission lines on a standard CMOS process. Their procedure to remove the substrate underneath the metal traces is as follows. First, areas on the passivation layer of the CMOS fabricated chips are open; this exposes the silicon substrate for postfabrication etching. Second, the ground/signal/ground (GSG) conductor strips are laid out in the first-layer metal (aluminum). Third, the wafers are subjected to a two-step etching process. In the first step, a gaseous isotropic etchant, xenon difluoride (XeF_2), is applied [17] to create small cavities around each open area. Continued exposure to this isotropic etchant for a period of 16 min leads to the connection of the etched cavities being formed beneath the two sides of the signal line. Once this point is reached, the second etch step begins. It consists of the immersion of the wafer into the anisotropic etchant EDP for about 1 hr at 92°C. The EDP etches the wafer bulk anisotropically to a cavity shaped by the $<100>$ crystalline structure of the silicon wafer, which results in a V-shaped pit, as shown in Figure 5.20.

The fabricated transmission lines are shown in Figure 5.21, and the measured performance, before and after etching, is shown in Figures 5.22 and 5.23.

To assess the effectiveness of eliminating the substrate from underneath the silicon lines, measurements of the transmission line insertion loss and effective dielectric constant were conducted. It was found that at 20 GHz, the insertion loss is decreased 24 dB, while at 40 GHz, a 34-dB improvement was registered. This improvement, of course, is not just due to the elimination of substrate losses; there is a reduction in metallic ohmic losses because keeping a certain characteristic impedance, after going to an air environment, entails a wider line. Improvements in dispersion characteristics, as well as a higher phase velocity, do indicate clearly, however, the advantages of eliminating the lossy silicon substrate. While it is a useful technique to interface microwave-like circuits with on-chip functions, there are some limitations to the performance obtainable using postprocessing bulk micromachining. In particular, the high-

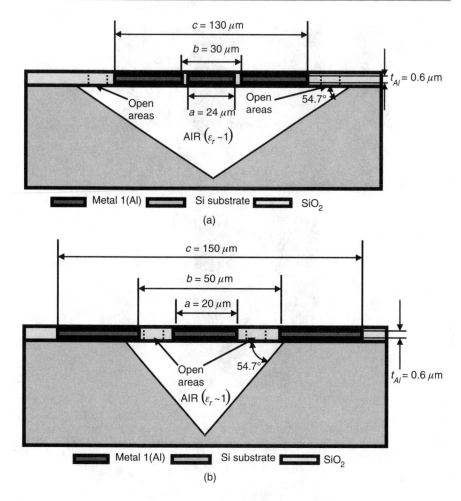

Figure 5.20 Cross-sectional view of the transmission-line structures with the etched V-shaped pit. (a) Open areas outside the ground planes, and (b) open areas between signal and ground planes. (*From:* [18], © 1997 IEEE. Reprinted with permission.)

density circuit environment of the chip precludes the definition of thick lines. Thus it may be impossible to tailor their metal cross-sectional area to achieve a prescribed series impedance.

5.3.1.4 LIGA-Micromachined Planar Transmission Line

As described in Chapter 2, the LIGA fabrication process allows the creation of tall (10 μm to 1 μm) metal structures with steep (high-aspect ratio) sidewalls on an arbitrary substrate material. These easily-realized attributes of LIGA micromachined structures, as shown in Figure 5.24, have been investigated by

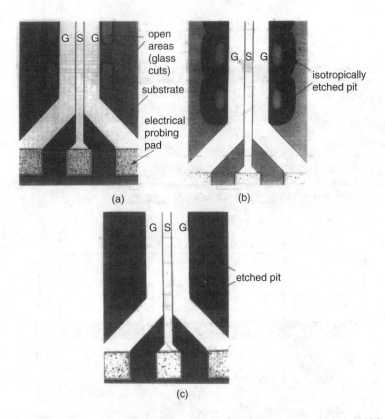

Figure 5.21 Microphotograph of the 50-Ω coplanar transmission line, (a) after CMOS fabrication, (b) after isotropic etch, and (c) after combined etch. (*From*: [18], © 1997 IEEE. Reprinted with permission.)

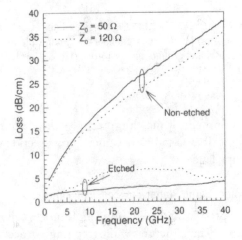

Figure 5.22 Attenuation measurements of the transmission lines before and after etching. (*From*: [17], © 1997 IEEE. Reprinted with permission.)

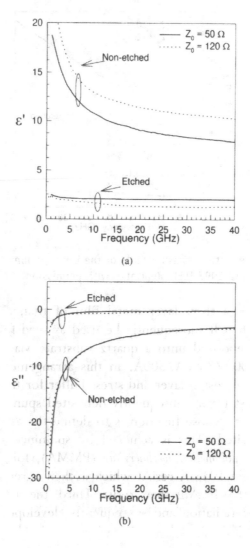

Figure 5.23 Measured effective dielectric constant of the transmission lines before and after etching: (a) real part, and (b) imaginary part. (*From*: [17], © 1997 IEEE. Reprinted with permission.)

Willke and Gearhart [19], in the context of applications requiring either very high levels of electromagnetic coupling, like certain passive microwave functions, or high-power handling capabilities, like monolithic transmitters.

The LIGA-micromachining fabrication process utilized by Willke and Gerhart [19] has the following steps, as shown in Figure 5.25. First, the metal substrate, a three-layer metal film of titanium, copper, and titanium, which will subsequently be used as seed for electroplating the mold, is deposited unto a

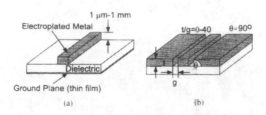

Figure 5.24 LIGA planar transmission-line geometry for (a) microstrip line, and (b) coplanar waveguide transmission line. (*From*: [19], © 1997 IEEE. Reprinted with permission.)

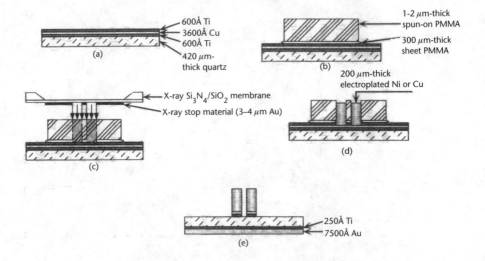

Figure 5.25 Cross sections of the LIGA fabrication process: (a) Application of a plating base, (b) application of sheet resist, (c) X-ray exposure of resist, (d) conductor electroplating, and (e) final structure. (*From*: [19], © 1997 IEEE. Reprinted with permission).

quartz substrate via sputtering, to a thickness of 600Å/ 3,600Å/600Å. In this arrangement, the titanium performs as an adhesion layer and stress buffer for the quartz, and the thick photoresist to be subsequently deposited spun on. Second, the thick photoresist, whose function is to define the aspect ratio of the mold, is deposited. This is achieved by spinning on a thin layer, 1 to 2 μm, of polymethyl methacrylate (PMMA), followed by a high molecular weight PMMA resist, which is then cured, and monomer-welded to the thin PMMA sheet resist. Third, the resist is exposed to coherent X-ray radiation and subsequently developed. Fourth, after development of the PMMA mold, the titanium capping layer is etched and the pattern-plating of nickel (nickel sulfamate), gold, or copper begins. Finally, the PMMA is dissolved and the three-layer metal film is etched away to electrically isolate the plated structures.

While no data on the loss performance of LIGA microstrip lines has been reported, analysis of LIGA microstrip lines on 420-μm-thick fused quartz (ε_r = 3.81 at 30 GHz) shows that a wide range of characteristic impedances are available, as shown in Figure 5.26 [19]. The line impedance decreases with increasing width, but decreases only slightly with increasing thickness.

5.3.1.5 Micromachined Waveguides

Despite the economies of scale and ease of manufacturability enabled by planar fabrication processes, the lowest microwave loss achievable in these technologies is ultimately limited by not only radiation losses, but by the unfavorable volume-to-surface area ratio (i.e., low Q), characteristic of thin planar structures. These inherent limitations of planar technologies leave system designers with no other choice but to resort to the utilization of waveguide components. Traditionally, however, the manufacturing of waveguide components has been the purview of conventional machining (metal-cutting) processes, which are based on materials such as brass, and which become difficult and expensive to apply as the dimensions of the components shrink (e.g., less than 0.3×0.15 mm for a 500- to 1,000-GHz waveguide). Since micromachining techniques are perfectly capable of defining 3D structures, the idea of applying them to make waveguide components in the context of a planar process, is rather appealing. Indeed, efforts along these lines have been pioneered by McGrath, et al. [20], who have applied bulk micromachining to fabricate rectangular waveguides for frequencies between 100 and 1,000 GHz on a silicon wafer.

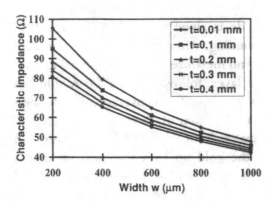

Figure 5.26 Characteristic impedance data for the LIGA microstrip line obtained via a finite-difference calculation. (*From*: [19], © 1997 IEEE. Reprinted with permission.)

The approach adopted by McGrath, et al. [20] to create a waveguide, and yet take advantage of planar assembly techniques, is to make two half sections split along the broadwall, as shown in Figure 5.27. Their process is as follows. They first deposit a 1,000Å thick layer of Si_3N_4 via LPCVD on both sides of the wafer. Second, they spin on photoresist, to be subsequently patterned with windows defining the "b" dimension of the guide. This is performed with a sulfur fluoride (SF_6) plasma. Third, they place the wafer in a reflux system and etch it in a water-based solution of 40% potassium hydroxide (KOH) at 80°C. This, the quintessential bulk micromachining step, defines steep walls by exploiting the difference in etch rates between the (110) and (111) planes of the silicon bulk, which is 170:1. Fourth, once the waveguide grooves have been defined, the half waveguide is glued to another identical 0.05-in-thick wafer via polyimide bonding. After this, they dice the wafer into pieces of half waveguides and metallized. The metallization process uses a 200Å chrome adhesion layer, followed by a 5,000Å gold layer using vacuum evaporation. To ensure uniform coverage, despite the steep walls, they performed the evaporation at three different angles, namely, 0°, and ±45°. Finally, from skin-depth considerations, they determined to electroplate the gold up to a thickness of approximately 3 μm, which is close to 12 times the skin depth at 100 GHz. This would reduce microwave ohmic losses.

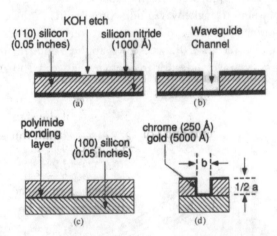

Figure 5.27 Cross-section view of the waveguide fabrication process: (a) Si_3N_4 mask defines the waveguide height; (b) wafer is etched completely through; (c) wafer with waveguide channels is bonded to an unetched wafer which forms the waveguide side wall; and (d) completed half-section of waveguide with gold plating. Two of these sections are mated to form the waveguide. "a" is the waveguide width and "b" is the height. (*From*: [20], © 1993 IEEE. Reprinted with permission.)

The measured insertion loss reported by McGrath et al., on waveguide fabricated according to the process just described, was of 0.04 dB per wavelength across the 75- to 110-GHz band. This is quite comparable to the 0.024 dB per wavelength value of commercially available metal waveguides, and thus validates their process and the potential for bulk-micromachining-based integrated "planar" waveguide circuit fabrication.

5.3.1.6 Micromachined Stripline

Stripline, as shown in Figure 5.28, is a transmission structure that may be thought of as a planar *coaxial* transmission line. In particular, the "center" conductor, a planar metal strip, is supported by a planar substrate, just like the microstrip line; but instead of having air above it, a dielectric layer is deposited on top of it, and this is subsequently metallized. As a result, the planar strip embedded in a dielectric and enclosed by a rectangular shield.

A typical application of striplines is in couplers, where, due to their shielded nature, radiation losses, and thus insertion losses, are minimized. A potential source of performance degradation, in the context of micromachining, is the nature of the dielectric utilized in its implementation. US-8 [21], an epoxy-based photoresist employed extensively to realize thick high aspect ratio MEMS structures, was used to prototype a micromachined stripline [22]. The fabrication procedure is sketched in Figure 5.29.

First, the ground plane was defined by sputtering 50 nm of TiW, 100 nm of Ni, and electroplating a 2.5-μm-thick copper layer, as shown in Figure 5.27(a). Second, the supporting dielectric layer, SU-8, was spin-casted and patterned to provide a mold for the subsequent electroplating of vias connecting the top metal cover to ground as shown in Figure 5.29(b). SU-8 dielectric deposition was then followed by planarization to a thickness of 16.5 μm, as shown in Figure 5.29(c). Third, copper electroplating was effected, followed by stripline and contact definition, as shown in Figure 5.29(d). Fourth, the top dielectric is spun-deposited, as shown in Figure 5.29(e); and fifth, the top metallization is deposited, as shown in Figure 5.29(f).

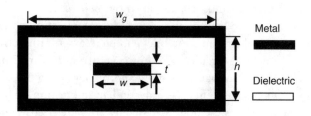

Figure 5.28 Sketch of stripline cross section.

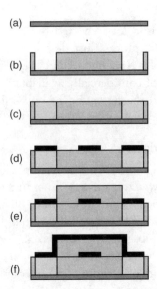

Figure 5.29 Fabrication steps for micromachined stripline: (a) Ground plane deposition, (b) supporting dielectric substrate, (c) electrodeposition and planarization of copper, (d) deposition and patterning of stripline, (e) deposition of top dielectric, and (f) deposition of top metallization. (*After:* [22].)

The micromachined stripline with geometrical parameters, w_g = 48 μm, h = 35.5 μm, w = 20 μm, t = 2.5 μm, resulting in a characteristic impedance of 48, exhibited insertion loss of 0.19 dB/mm and 0.58 dB/mm, at 10 GHz and 48 GHz, respectively. The extracted relative dielectric constant (ε_r) and loss tangent [tan (δ)] were 3.1 and 0.043, respectively, at 48 GHz.

5.3.2 Micromachining-Enhanced Lumped Elements

The utilization of transmission lines in microwave circuits and systems is motivated by the fact that as at these frequencies, component dimensions are comparable to the wavelength of the signals propagating through them, the currents and voltages in the circuit must be described by their *distributed* spatial profile. This is certainly bound to be the case for lines interconnecting nonadjacent circuits, as well as pieces of transmission lines chosen to exhibit inductive or capacitive behavior. On a local circuit scale, however, structures may be fabricated which are small enough, so that it is a good approximation to take their voltages and currents as *lumped* at a point (i.e., independent of position). The advantages of lumped-element circuits include their smaller size, lower cost, and wideband characteristics [11]. These features are particularly important in applications where yield is a concern (e.g., RFICs and MMICs), and for wideband applications. Moreover, lumped elements are more efficient at implementing impedance matching transformations, since they enable a greater range of ratios (e.g., for instance, 20:1), as compared to distributed-circuit matching [11]. These features make lumped elements the elements of choice when it comes to high-power oscillators, power amplifiers, and broadband circuit applications. Unfortunately, despite their small size, the performance of lumped microwave components is limited by the manifestation of parasitic effects, which change the intrinsic properties of the "pure" elements into those of a composite RLC circuit, which then exhibits resonance. In the case of an inductor, for example, its nature upon self-resonating will become capacitive above the resonance frequency, thus losing its usefulness. Since the major source of these parasitic effects is the dielectric substrate supporting the elements, it is natural to aim at improving the situation by eliminating it.

5.3.2.1 Lumped Inductors

In the context of a conventional planar process, a lumped inductor may be realized using a high-impedance (i.e., very narrow) microstrip section or a spiral conductor as shown in Figure 5.30.

The innermost turn of the spiral inductor is connected to the other circuitry through a conductor that passes under an airbridge in RFICs and MMICs, whereas a wire bond connection is made in MICs. In this structure, the width and thickness of the conductor under the airbridges determine the current-carrying capacity of the inductor. They possess typical thickness of 0.5 μm to 1.0 μm, and the airbridge separates it from the upper conductors by 1.5 μm to 3.0 μm [11]. In RFICs and MMICs operating above 1 GHz, the typical inductance values utilized range from 0.5 to 10 nH. As the inductance of the spiral is increased, so does the capacitance to the substrate. This behavior leads to the simultaneous reduction of the inductor's self-resonance frequency. For example, while gold spiral inductors of 25 nH are found to self-resonate at about

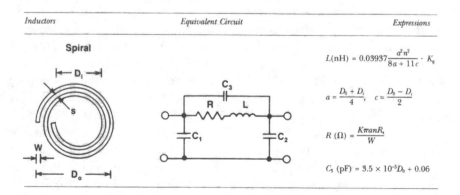

Inductors	Equivalent Circuit	Expressions

$$L(\text{nH}) = 0.03937 \frac{a^2 n^2}{8a + 11c} \cdot K_g$$

$$a = \frac{D_0 + D_i}{4}, \quad c = \frac{D_0 - D_i}{2}$$

$$R\ (\Omega) = \frac{K\pi a n R_s}{W}$$

$$C_3\ (\text{pF}) = 3.5 \times 10^{-5} D_0 + 0.06$$

Figure 5.30 Spiral inductor [12].

3 GHz on GaAs substrates, aluminum inductors only as large as 10 nH on standard silicon substrates will self resonate at 2 GHz. In addition to the self-resonance frequency, the Q of an inductor is essential to its usefulness in applications such as low-loss VCOs, low-loss impedance matching networks, low-loss filters, and low-noise amplifiers. In particular, at frequencies below their self-resonance frequency, the factors determining the Q of inductors may be gauged from the expression [23]:

$$Q \approx \frac{\omega L}{R_S(\omega)} \cdot \left(substrate\ loss\ factor\right) \cdot \left(self - resonance\ factor\right) \quad (5.4)$$

where $R_s(\omega)$ is the series resistance. Accordingly, the Q may be maximized by reducing the series resistance, by eliminating the losses to the substrate (i.e., making the substrate loss factor $\rightarrow 1$, and by operating well below self-resonance (i.e., making the self-resonanct factor $\rightarrow 1$).

A number of efforts have been undertaken to improve the Q of planar inductors [24], from schemes aimed at reducing eddy current-induced losses, due to magnetic field penetration into both the substrate and adjacent traces; to techniques aimed at reducing trace resistance, which originates in its finite conductivity and width; to reducing substrate capacitance and ohmic losses, which shunt the input signal to ground and thus cause less than the input signal to reach the output. However, the typical Q resulting from these approaches is seldom greater than 10 for a 1 nH, in the 1- to 4-GHz frequency range [24]. This overt limitation, embodied in the ever-present spreading resistance of semiconductor substrates, justifies its elimination [23].

A more precise assessment of the effect of the substrate parasitics may be obtained from a look at the approximate expressions for inductances and resistances in spiral inductors, given in Figure 5.30 [11]. In the model, n is the number of turns, S is the spacing between turns, R_s is the sheet resistance of the conductor per square, l is the length of the conductor, and K_g is a correction factor that takes into account the decrease in inductance value as the substrate thickness is decreased. A closed-form expression for K_g for a ribbon is given by $K_g = 0.57 - 0.145l \cdot nW/h$, $W/h > 0.05$, where W is the conductor width and h is the substrate thickness [11]. C_1 and C_2 represent the parasitic capacitance between the spiral inductor top metal and the bottom ground plane. This capacitance value depends on the dielectric constant of the substrate. Since C_1, $C_1, C_2 \sim \varepsilon_r$, etching away the dielectric material between the spiral inductor and the ground plane would reduce $C_1, C_2 \sim \varepsilon_r$ by a factor of ε_r.

Now consider micromachining. The top-side etching micromachining technique discussed in Section 5.3.2 was successfully applied by Chang, et al. [25], to remove substrate thicknesses between 200 μm and 500 μm, resulting in a measured inductance value of 114 nH with an accompanying increase in self-resonance frequency from approximately 800 MHz, before etching away the silicon substrate, to 3 GHz, after substrate removal. Other efforts along these lines include the following. Lopez-Villegas, et al. [26], in addition to removing the substrate from underneath the spiral, laid out a nonuniform diminishing trace width to minimize eddy current losses near the center of a spiral inductor, where the field is highest and the losses greatest. They obtained a Q of 18 at 5 GHz on an 8 nH inductor, and a Q of 15 in the 1- to 2-GHz range on a 34-nH inductor, with a 3-μm-thick Au trace. Jiang, et al. [27], demonstrated inductors suspended over deep copper-lined cavities. The spirals consisted of copper-encapsulated polysilicon, and typical performance included a Q of 36 at 5 GHz on a 2.7-nH inductor. Yoon, et al. [28], demonstrate surface micromachined air-core solenoidal inductors on silicon with a Q of 16.7 at 2.4 GHz on a 2.67-nH inductor. Similarly, air-core inductors configured as toroids have been demonstrated by Liu, et al. [29], on a low resistivity silicon wafer with a Q of 22 at 1.5 GHz, and a self-resonance frequency greater than 10 GHz on a 2.45-nH inductor.

In a new direction to minimize the effects of substrate parasitics, Dahlmann, et al. [30], exploited the surface tension of solder to effect the self-assembly of inductors disposed perpendicularly to the wafer plane, obtaining a Q of 30 at 3 GHz on a 2-nH meander inductor. Lubecke, et al. [31], demonstrated a self-assembly approach, in which the interlayer stress between metal and polysilicon causes the patterned inductors to curl up away from the substrate upon release, as shown in Figure 5.31. A Q of 7 at 2 GHz on a 1.5-nH inductor was obtained.

Along the lines of this same theme, Chen, et al. [32], exploited a microstructure assembly technique called plastic deformation magnetic assembly (PDMA), in which covering the inductor traces with a magnetic material and turning on a bias magnetic field perpendicular to the wafer plane causes a torque that rotates the structure away from the substrate (see Section 2.1.4). Using this technique, a Q between 5 and 10 in the 1- to 4-GHz range on a 4.5-nH inductor was demonstrated.

Further examination of inductor performance in planar processes, particularly aimed at optimizing the first term of (5.4) for general planar processes concluded that there is a linear dependence between the inductor area and its Q, namely [23],

$$Q_{\max-2D} \approx \frac{1}{8} D_{out} / \delta_{skin} \qquad (5.5)$$

where D_{out} is the outer diameter of the spiral, and σ_{skin} is the skin depth. Accordingly, obtaining a Q of 50 at 1 GHz requires an area of 1mm^2. A similar

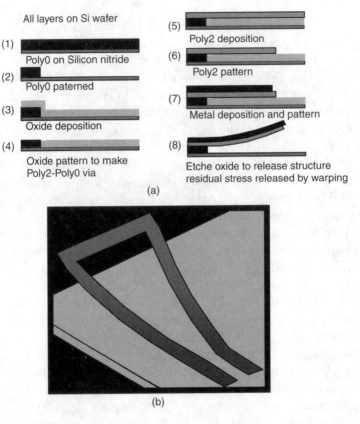

Figure 5.31 (a) Simplified fabrication process for self-assembling inductors. and (b) scanning electron micrograph (SEM) of triangular self-assembling inductor showing inductor curling away from the substrate. (*After:* [31], 2001 IEEE. Courtesy of Dr. V. M. Lubecke, University of Hawaii.)

study, for the case of a three-dimensional *toroidal* inductor, concluded that for an N-turn torus of radius r_{torus} and winding radius r_{coil} of inductance

$$L = \frac{\mu_0 \mu_r N^2 r_{torus}^2}{2r_{coil}}$$

(5.6)

where

$$r_{coil} = \frac{D_{out}}{2} - r_{torus},$$

(5.7)

the maximum Q is given by:

$$Q_{max-3D} = \sqrt{\frac{\mu_0 \cdot \mu_r}{\pi \cdot \rho}} \frac{\sqrt{f} \cdot r_{torus} \cdot \left(2\pi \cdot r_{coil} - N \cdot s_{coil}\right)}{\alpha(f) \cdot r_{coil}} \tag{5.8}$$

where μ_0 is the permeability of vacuum, μ_r is the relative permeability of the toroidal core, ρ is the coil resistivity, $\alpha(f)$ is a coefficient that takes into account current crowding effects (the effect causing changes in the internal current distribution of conductors when they are in close proximity), and s_{coil} is the separation between the turns of a coil. Armed with the results of (5.5) and (5.8), a quantitative comparison between the Q of planar and toroidal inductors, for a given inductance value, frequency, and outer diameter, results in a Q of the toroidal inductor that exceeds that of the planar spiral by a factor of two. In particular, assuming gold conductors, $D_{out} = 1.5$ mm, and $r_{torus} = 315$ μm, a Q of 180 for the toroidal inductor is obtained at 1 GHz. The concept for the realization of the toroidal inductor is shown in Figure 5.32 [23].

The design equations for a toroidal inductor, whose equivalent circuit model in the frequency range between dc and 10 GHz is given in Figure 5.33, and whose inductance and Q are given in (5.6) to (5.8), are as follows [23]:

$$R_{series} \approx R_{dc}\left[1 + \left(\frac{f}{f_{cr}}\right)^{a^1 + a^2}\right] \tag{5.9}$$

(a)   (b)

Figure 5.32 Visualization of method of fabrication for toroidal inductor: (a) Horizontal split of the toroidal inductor, and (b) joining of two halves to achive final geometry. (*Source:* [23], © 2003 IEEE. Courtesy of Mr. H. Nieminen, Nokia Research Center, Helsinki, Finland.)

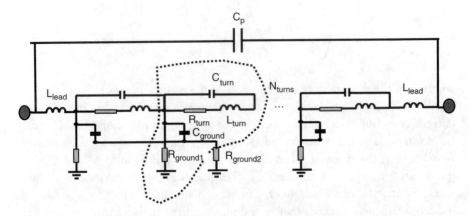

Figure 5.33 Equivalent circuit of the toroidal inductor. C_{ground} is the capacitance of a single coil turn to ground, $R_{ground1}$ is the ohmic substrate loss per turn, $R_{ground2}$ is the ohmic loss in series with the substrate capacitance, L_{lead} is the parasitic lead inductance, and $C_p = C_{turn}$ captures the coupling between the first and last turns of the coil. (*After* [23].)

$$R_{dc} = \frac{2\pi r_{torus} N^2 \rho}{2\pi r_{coil} - N \cdot s_{coil}} \tag{5.10}$$

$$R_{turn} = \frac{R_{series}}{N} \tag{5.11}$$

$$C_{ground} = \frac{2\pi \cdot r_{coil} \cdot \varepsilon}{59.952 \cdot N \cdot V_0 \cdot \ln\left(\frac{2 \cdot h}{D_{torus}} + \sqrt{\left(\frac{2 \cdot h}{D_{torus}}\right)^2 - 1}\right)} \tag{5.12}$$

where R_{dc} is the total series resistance of the coil at dc, f_{cr} is the critical frequency at which the series resistance begins to be dominated by the skin depth, α_1 and α_2 are coefficients related to skin depth and current crowding, respectively, ε is the substrate permittivity, h is its thickness, V_0 is the speed of light, and D_{torus} is the diameter of the torus cross section.

The fabrication steps for the toroidal inductor are depicted in Figure 5.34. The concept involves the fabrication of two halves (see Figure 5.32), followed by their integration using thermocompression bonding and ion milling for opening the access to the contacts. An important consideration was the utilization of a method of replication to simultaneously achieve the accuracy and small features germane to silicon micromachining, together with the economies of scale characteristic of the data storage industry. In this context, a master was first

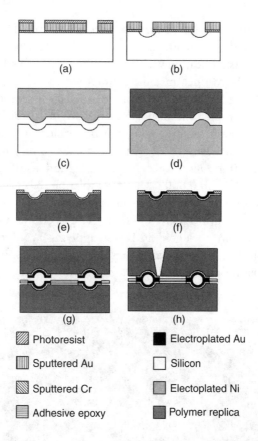

(a)	(b)
(c)	(d)
(e)	(f)
(g)	(h)

▨ Photoresist   ■ Electroplated Au

▥ Sputtered Au   ☐ Silicon

▨ Sputtered Cr   ▥ Electoplated Ni

▤ Adhesive epoxy   ■ Polymer replica

Figure 5.34 Manufacturing process for the toroidal inductor. (*From*: [23], © 2003 IEEE. Reprinted with permission. Courtesy of Mr. H. Nieminen, Nokia Research Center, Helsinki, Finland.)

produced in silicon or glass with the desired geometry, followed by the electroplating of a negative copy of a master to create mould inserts. The mould inserts could then be replicated via injection molding, casting, or hot embossing of a polymer to produce large quantities. Due to the fact that the dielectric properties of cyclo-olefin polymer, a thermoplastic polymer, are superior to those of thermosetting materials, and that it is easier to obtain flat and stress-free wafers by injection molding in comparison to casting of thermosetting polymers, injection molding of COP was employed. A few details on the fabrication of the master were given. Their production utilized the isotropic etching of silicon. This entailed avoiding stirring and disposing the wafer horizontally, while the features were oriented upwards. Next, an etching solution consisting of proportions of 96:4 or 91:9 HNO_3 (69%): HF (50%) producing an etching rate of 1 μm per min at room temperature, was devised. Thus, by avoiding acetic acid, it was found that smooth polished surfaces were obtainable. For a 120-μm etch depth, the isotropy (width:depth) obtained was 1:0.96. Once the master creation was complete, its replication was achieved by transferring the silicon master structure into a 300-μm-thick metallic counterpart by electroplating, a process that involved plating nickel from a sulfamate electrolyte. At this point, the mould obtained resembled the negative of the original structure, as shown in Figure 5.34(c). Injection molding was then performed on two different polymers, namely, polycarbonate (PC), which has a dielectric constant and a loss tangent of 3.5 and 0.01, respectively, at 1 MHz, and COP, which has corresponding values of 2.3 and 0.0002. Upon completing the creation of the toroid halves on these polymers, they were metallized with the traces of the coils and the electrodes, as shown in Figure 5.34(e, f). The electrodes were patterned by first sputtering a 50-nm-thick gold seed layer, applying a conformal PR, namely, Shipley PEPR 2400, onto the polymer substrates, and developing the PR, subsequently electroplating pure gold (from a potassium gold cyanide system) to a

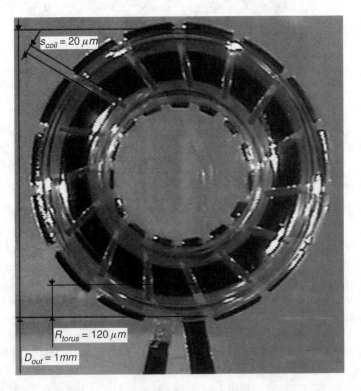

Figure 5.35 Light optical microscope photograph of an assembled inductor. Dimensions: r_{torus} = 120 μm, D_{out} = 1 mm, s_{coil} = 20 μm, metal thickness 4 μm. (*From*: [23], © 2004 IEEE. Reprinted with permission. Courtesy of Mr. H. Nieminen, Nokia Research Center, Helsinki, Finland.)

thickness of 4 to 6 μm Following electroplating, the gold seed layer was etched with HCl:HNO$_3$:H$_2$O in proportions 4:4:9 to isolate the structures. The final step entailed joining the two halves by thermocompression, with the heat-curing adhesive EpoTeck OG198-50 at a temperature and pressure of 120°C and 75-bar, respectively, for 15 min. Figure 5.35 shows the assembled toroidal inductor.

The measured toroidal inductor performance was an inductance of 6 nH exhibited a maximum Q of 50 at 3 GHz, and a temperature dependence of 30 ppm/K.

5.3.2.2 Lumped Capacitors

Two types of types of passive capacitors are generally used in microwave circuits, namely, the interdigital and the metal-insulator-metal (MIM) capacitors, Figure 5.36. The application of micromachining to the interdigital capacitor is the most obvious, since its main function is to implement a *series* (e.g., blocking capacitor), and has received the most attention. Since capacitance is proportional to the occupied area, small capacitance values (e.g., of the order of 1 pF and below) are realized with interdigital capacitors, while higher values invoke MIM techniques to minimize the overall size [12].

Chi [14] has also investigated the application of micromachining to membrane-supported planar interdigital capacitors. In this case, he found that the planar capacitors do not suffer from a low resonant frequency, but rather from a relatively large shunt parasitic capacitance to ground C_1 [11]. The micromachining membrane approach is less effective in this case, because, while it reduces the parasitic capacitance C_1 by a factor of ε_r, it also reduces the interdigital series capacitance by a factor of $(1 + \varepsilon)/2$ [11]. Nevertheless, Chi concluded that the microwave performance of interdigital capacitors fabricated on a membrane is superior to that obtained from those fabricated on high resistivity silicon substrate [14]. Specifically, they should be expected to exhibit a larger quality factor due to the elimination of the substrate's dielectric losses.

Finally, by removing the substrate bulk from underneath the bottom plate of MIM capacitors, Sun, et al. [33], demonstrated that at 2 GHz, the Q of a

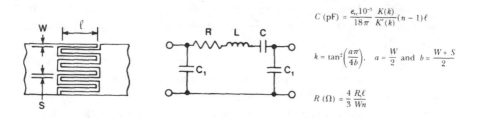

$$C \text{ (pF)} = \frac{\epsilon_{re}10^{-3}}{18\pi} \frac{K(k)}{K'(k)}(n-1)\ell$$

$$k = \tan^2\left(\frac{a\pi}{4b}\right), \quad a = \frac{W}{2} \text{ and } b = \frac{W+S}{2}$$

$$R \text{ (}\Omega\text{)} = \frac{4}{3} \frac{R\ell}{Wn}$$

Figure 5.36 Lumped capacitors [11].

2.6-pF MIM capacitor could be improved from less than 10, before substrate removal, to greater than 100 (together with a self-resonance frequency of 15.9 GHz), after substrate removal.

5.3.2.3 Tunable Capacitors

While many microwave circuit functions require variable capacitors [e.g., voltage-controlled oscillators (VCOs)], the progress towards their full monolithic integration has been traditionally hampered by the difficulty of realizing varactors in monolithic form [25]. Besides concerns stemming from process incompatibility, limitations imposed by excessive series resistance, low quality factor, and limited tuning range top the list of offenders [34]. Recently, however, there has been vigorous activity aimed at exploiting MEMS fabrication techniques to realize on-chip varactors.

For example, in [35], Dec and Suyama proposed and demonstrated a three-plate capacitor. The structure consists of a grounded movable metal plate, which is sandwiched between two fixed metal plates. As the voltage between the grounded movable plate and either of the fixed plates is varied, the movable-fixed plate capacitance changes. A prototype device, fabricated in a polysilicon surface micromachining process, exhibited a tuning range of 25% and a quality factor of 9.6 at 1 GHz when the capacitance is tuned to 4 pF.

In another approach, Young and Boser [36] proposed an electrostatically-actuated aluminum plate suspended over a fixed bottom plate fabricated in a CMOS-compatible process. They point out that their choice of aluminum instead of polysilicon as the structural material was dictated by two key reasons, namely, its lower sheet resistance, which is critical to minimize ohmic losses and to guarantee an adequate quality factor, and its low processing temperature of only 150°C. Physically, the device consists of two suspended plates with a dc bias voltage applied across the capacitor. As this is essentially a cantilever beam structure, it is necessary to avoid entering the bias voltage regime of spontaneous deflection. Thus the applied bias is limited to values prescribing nonhysteretic actuation. This immediately may be expected to impose a limitation on the maximum obtainable tuning range. For example, a prototype device tested at 1 GHz, consisting of four parallel microstructures, is tunable between 2.11 and 2.46 pF with a 5.5 dc tuning voltage range. The quality factor measured was masked by the resistive interconnect losses and was reported to be 62 [36].

In an attempt to overcome the tuning range limitation germane to parallel-plate type variable capacitors, Hung and Senturia proposed tunable capacitors with programmable capacitance-voltage characteristics [37]. This device, called *zipping* actuator, departs from conventional approaches in two fundamental respects: it exploits the whole actuation regime (i.e., before and after the spontaneous deflection), and it uses a *shaped* bottom electrode. As a result, further increase in bias voltage after beam collapse, causes simultaneous propagation of the diminishing beam-to-substrate gap distance as the tip flattens towards the anchor, as well as a variation in the overlap area, thus "zipper." The capacitance between the beam and the bottom electrode is, therefore, controlled by the applied dc bias voltage, and the C-V characteristic of the device is determined by the *geometry* of the bottom electrode. Cantilever dimples in this approach serve two purposes, namely, they hold the beam at a small distance from the bottom electrode to prevent a short circuit, and through their depth, control the capacitive tuning range. Prototype devices of the zipping actuator, implemented in the MUMPs process, achieved a 25% tuning range in the zipping regime, with a linear C-V characteristic [37].

Further work on the zipping-based varactor concept was subsequently undertaken by Ionis, Dec, and Suyama [38], who implemented a varactor in the MUMPs polysilicon surface micromachining process, and demonstrated a 46% (3.12 to 4.6 pF) tuning range with control voltage of 0 to 35V, while exhibiting a Q of 6.5 at 1.5 GHz.

Using a vertical thermal actuator to separate the top plate from the bottom fixed plate, Feng, et al. [39], demonstrated a varactor with a tuning range of 270%, and a Q of 300 for a 0.1-pF nominal capacitance at 10 GHz. The prototype was implemented by combining MUMPs and flip-chip assembly together with a substrate transfer technique. On the other hand, using Taiwan Semiconductor Manufacturing Company's (TSMC) 0.35 μm CMOS foundry process, Oz and Fedder [40] demonstrated an electrothermal RF CMOS-MEMS varactor exhibiting a 352% tuning range (42 to 148 fF) with a 0- to −12-V control voltage range, and a power dissipation of 34 mW at 1.5 GHz.

Another approach to overcome the electrostatic pull-in phenomena limiting parallel-plate varactor tuning range, was demonstrated by Chen, et al. [32], as shown in Figure 5.32. In this approach, the functions of actuation electrode and varactor are decoupled, as shown in Figure 5.37(a). Its advantage may be gauged from the formula giving the tuning range, namely:

$$\frac{C - C_0}{C_0} = \frac{\varepsilon A/(h_1 - h) - \varepsilon A/h_1}{\varepsilon A/h_1} = \frac{h}{h_1 - h} \tag{5.13}$$

where h is the top plate displacement upon actuation. In particular, for the conventional parallel-plate varactor, the maximum displacement, set by pull-in, is $h = h_1/3$, which substituted in (5.13) gives a maximum tuning range of 50%. For the new varactor, making $h_1 \leq h_2/3$, the pull-in limit is never reached, and consequently h may approach h_1 very closely, making the potential tuning range large. For example, with $h_1 = 2\,\mu$m and $h_3 = 3\,\mu$m, a displacement $h = h_2/3 = 1$ μm yields an ideal tuning range of 100%. The fabrication sequence for this

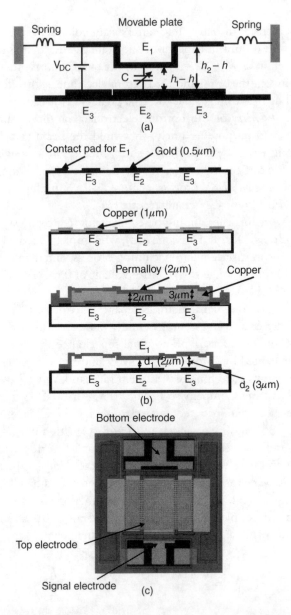

Figure 5.37 (a) Sketch of wide tuning range varactor, and (b) sketch of fabrication process flow for wide-tuning range varactor. (*After:* [32].) (c) Top view of two-gap varactor implementation. (*From:* [41]. Courtesy of Mr. H. Nieminen, Nokia Research Center, Helsinki, Finland.)

device is interesting, in that it exploits and engineers the sacrificial layer topography to achieve a variable height. In particular, the fabrication begins by thermally evaporating and patterning a 0.5-μm-thick gold layer to define the fixed electrodes E_2 and E_3, and the pads for contacting top electrode E_1 on a Pyrex glass wafer. Next, the variable height sacrificial layer is deposited in two steps. First, an initial layer of copper is thermally evaporated to a 1 m thickness, and patterned. In particular, the copper deposited over electrode E_2, the bottom plate (signal electrode) of the varactor, is etched away. Then, another 2-μm-thick copper layer is thermally evaporated. This results in a sacrificial layer thickness that is 1 μm thinner over E_2 than over E_3 (the actuation electrodes); thus, it possesses a variable height. Following copper sacrificial layer deposition, the top plate is deposited by using it as a seed layer for electroplating a 2-μm-thick layer of Ni-Fe alloy. The last step is copper sacrificial layer etching with a copper etchant consisting of $HAC:H_2O_2:H_2O$ in proportions 1:1:10 (where HAC stands for acetic acid, chemical formula: CH_3COOH), and drying with supercritical carbon dioxide (CO_2). The measured tuning range of varactor samples, with h_1 = 2 μm and h_3 = 3 μm, was between 50.9% and 69.8% at 1 MHz. Factors, such as surface roughness, fringing and parasitic capacitance in connection with the measurement systems, and electrode curvatures, were identified as potential causes for this limited range.

A subsequent implementation of the variable-gap varactor concept was undertaken by Nieminen, et al. [41], at Nokia, utilizing Tronic's Microsystems foundry services. They demonstrated a varactor, with h_1 = 0.5 μm and h_3 = 1.5 μm, that exhibited a tuning range of 270% and a Q of 66 at 1 GHz, and 53 at 2 GHz, at a nominal capacitance of 1.15 pF. The control voltage range was 0 to 17.7V. In another independent implementation of this concept, with h_1 = 0.75 μm and h_3 = 2.75 μm, Tsang, et al. [42], demonstrated a maximum tuning range of 433%.

5.3.2.4 Tunable Inductors

The proliferation of wireless standards, and the desirability to increase the functionality and flexibility of portable wireless appliances to accommodate these, has elicited great interest in the development of reconfigurable and programmable circuits. Until recently, the emphasis to achieve reconfigurability was focused on utilizing varactors, which tend to exhibit higher Qs than inductors. Recently, however, reports on efforts aimed at developing MEMS-based variable inductors (variometers) have begun to appear in the literature [43–46]. Many benefits of simultaneously possessing varactors and variometers at the circuit designer's disposal have been proposed by Fischer, Eckl, and Kaminski [43]. In particular, their concept of a universal matching network, as shown in Figure 5.38 [43], makes explicit use of both varactors and variometers, in addition to switches, to achieve wide tuning. Implemented in the context of matching networks, low-noise and power amplifiers, filters, VCOs, and so on, these would be key enablers of reconfigurable multiband and multistandard radios.

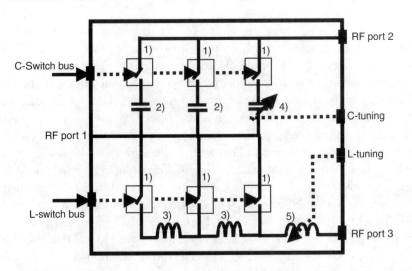

Figure 5.38 Universal matching network with wide tuning. (*After* [43].)

Various approaches to variometers have been advanced. For instance, Zine-El-Abidine, Okoniewski, and McRory [44], [see Figure 5.39(a)], employ the proximity of two coils connected in parallel which changed by actuating one of them with a thermal actuator that exerts a force as shown and causes it to buckle and lift up. This changes the mutual inductance between the two inductors and, thus the inductance seen at the terminals. A tuning range of 13% (1.045 to 1.185 nH) in the 2- to –5-GHz frequency range was reported.

In a somewhat similar inductance tuning concept, demonstrated by Tassetti, Lissorgues, and Gilles [45], the effective magnetic permeability of a primary inductor is tuned by varying its coupling to an array of metallic loops. The loops embody an artificial magnetic layer and provide a vertical magnetic flux that threads the primary inductor, thus effecting a magnetic coupling. In a proof-of-concept implementation, as shown in Figure 5.39(b), the primary inductor is fabricated on a substrate and the loop is fabricated in a cantilever beam. Then, by actuating and deflecting the beam, the loop proximity to the inductor underneath is changed, thus changing the magnetic coupling, and consequently the inductance at the terminals of the primary inductor, as its equivalent input impedance obeys,

$$Z_{eq}(\omega) = \left(Z_1 + j\omega L_1\right) + \frac{\omega^2 M^2}{\left(Z_2 + j\omega L_2\right)} \qquad (5.14)$$

where L_1, L_2, and M are the primary, secondary (loop), and mutual inductances, respectively. Z_1 is a parasitic impedance accompanying Z_1, and Z_2 is an

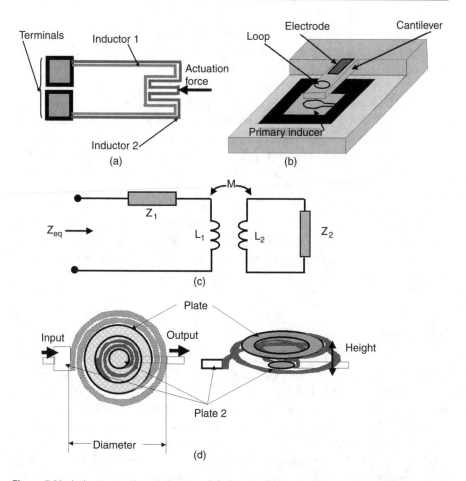

Figure 5.39 Inductor tuning schemes: (a) By exploiting buckling to separate parallel inductors, and (b) by exploiting cantilever beam deflection to vary magnetic coupling of inductor patterned on beam and artificial magnetic ground plane. (c) Equivalent circuit model for both (a) and (b). (d) unwinding spiral inductor.

impedance terminating Z_2. A tuning range of 36% (0.97 to 0.71 nH) in the 1- to 5-GHz range was achieved, with a 0- to –150-V actuation voltage range, for a primary inductor with a radius of 1.5 mm and metal trace width of 50 μm, and a 5×5 array of 300-μm radius loops and 50 μm similar trace width.

A third approach to variometers is shown in Figure 5.39(d) [46]. In this case, the effective distance between turns is varied by "unwinding" the inductor spiral. To accomplish this, a spiral structure is patterned with two plates—Plate 1, which is later pulled up; and Plate 2, which remains on the substrate's surface. To prepare Plate 1 for pull-up, a glass plate is bonded to it with a polyimide adhesive. Then, the glass plate is pulled up so that the spiral inductor becomes a three-dimensional stricture. At this point, the inductor is heated up until surpassing the glass transition temperature of 620K at 10^4 Pa. This is followed by thermal annealing, to release the internal stress in the spiral, and dissolution of the polyimide adhesive, to free the structure. The end result is a spiral with a height of 150 μm. Using an actuator to push down Plate 1, the inductance stretches, the interwinding distance increases, and the inductance changes. It was pointed out by Yokoyama, et al. [46], that the elasticity properties of the metal material are critical for this approach. They utilized $Pd_{76}Cu_7Si_{17}$, instead of Al or Cu, because it is highly flexible. In particular, its properties are as follows: Young's modulus, 57.9 GPa (tensile test)/69.7 GPa (bending test); tensile strength, 1.14 GPa; elastic limit, 1.97%; hardness, HV 515; density 10.4 × 10^3 kg/m^3; resistivity, 62.0 $\mu\Omega$-cm. A tuning range of 3% (3.64 to 3.74 nH) was obtained, together with a Q of 2 at 2 GHz, for an inductor with a diameter of 885 μm, and a height of 150 μm, patterned on a SiO_2 substrate and actuated by manually pushing with a teflon (ε_r = 2.2) block.

5.4 Summary

In this chapter, we have dealt with microwave MEMS applications. In particular, we have concerned ourselves with the subject of how MEMS technology can be exploited to enhance the microwave performance of fundamental devices. Starting with the switching function, we discussed in detail the operation, fabrication, and performance of a wide variety of implementations. Next, we considered a number of components enhanced by micromachining, namely, microstrip, coplanar, and waveguide transmission media, lumped fixed-value inductors and capacitors, and tunable capacitors and inductors. It is clear that imagination is the limit, as far as the exploits to which MEMS technology and techniques can be applied to great advantage in microwave circuits and systems, so that great improvements in performance at the device level will positively manifest themselves at the system level. An examination of these potentialities is the subject of next chapter.

References

[1]   Bacher, W., W. Menz, and J. Mohr, "The LIGA Technique and Its Potential for Microsystems—A Survey," *IEEE Trans. Ind. Electronics,* Vol. 42, 1995, pp. 431–441.

[2]   Frazier, A. B., R. O. Warrington, and C. Friedrich, "The Miniaturization Technologies: Past, Present, and Future," *IEEE Trans. Ind. Electronics,* Vol. 42, 1995, pp. 423–431.

[3]   Mehregany, M., "An Overview of Microelectromechanical Systems," *Proc. SPIE,* Vol. 1793, Integrated Optics and Microstructures, 1992, pp. 2–11.

[4] Sniegowski, J. J., and E. J. Garcia, "Microfabricated Actuators and Their Applications to Optics," *Proc. SPIE*, 1995, pp. 1–19.

[5] Bryzek, J., K. Petersen, and W. McCulley, "Micromachines on the March," *IEEE Spectrum*, 1994, pp. 20–31.

[6] de los Santos, H. J., et al., "Microwave and Mechanical Considerations in the Design of MEM Switches for Aerospace Applications," *Proc. IEEE Aerospace Conference*, Aspen, CO, February 1–8, 1997, pp. 235–254.

[7] Zavracky, P. M., S. Majumder, and N. E. McGruer, "Micromechanical Switches Fabricated Using Nickel Surface Micromachining," *J. Microelectromechanical Syst.*, Vol. 6, 1997, pp. 3–9.

[8] Yao, J. J., and M. F. Chang, "A Surface Micromachined Miniature Switch for Telecommunications Applications with Signal Frequencies from DC up to 4 GHz," *8th Int. Conf. on Solid State Sensors and Actuators, and Eurosensors IX*, 1995, pp. 384–387.

[9] Robert, P., et. al., "Integrated RF-MEMS Switch Based on a Combination of Thermal and Electrostatic Actuation," *Transducers 2003*, Boston, MA, June 8–12, 2003, pp. 1714–1717.

[10] Ruan, M., J. Shen, and C. B. Wheeler, "Latching Micromagnetic Relays," *J. Microelectromechanical Syst.*, Vol. 10, No. 4, December 2001, pp. 511–517.

[11] Gupta, K. C., et al., *Microstrip Lines and Slotlines*, Norwood, MA: Artech House, 1996.

[12] Ramo, S., J. R. Whinnery, and T. Van Duzer, *Fields and Waves in Communication Electronics*, New York: John Wiley & Sons, Inc, 1984.

[13] Weller, T. M., et al., "New Results Using Membrane-Supported Circuits: A Ka-Band Power Amplifier and Survivability Testing," *IEEE Trans. Microwave Theory Tech.*, Vol. 44, 1996, pp. 1603–1606.

[14] Chi, C.-Y., "Planar Microwave and Millimeter-Wave Components Using Micromachining Technologies," Ph.D. dissertation, Radiation Laboratory, University of Michigan, 1995.

[15] Weller, T. M., L. P. B. Katehi, and G. M. Rebeiz, "High Performance Microshield Line Components," *IEEE Trans. Microwave Theory Tech.*, Vol. 44, 1995, pp. 534–543.

[16] Milovanovic, B., V. Markovic, and D. Stojkovic, "High Frequency Circuits Manufactured by Micromachining Techniques," *Proc. 21st International Conf. Microelectronics*, Nish, Yugoslavia, September 14–17, 1997, pp. 523–526.

[17] Chang, F. I.-J., et al., "Gas-Phase Silicon Micromachining with Xenon Difluoride," *SPIE 1995 Symp. Micromach. Microfab.*, Austin, TX, October 1995, pp. 117–128.

[18] Milanovic, V., et al., "Micromachined Microwave Transmission Lines in CMOS Technology," *IEEE Trans. Microwave Theory Tech.*, Vol. 45, 1997, pp. 630–635.

[19] Wilke, T. L., and S. S. Gearhart, "LIGA Micromachined Planar Transmission Lines and Filters," *IEEE Trans. Microwave Theory Tech.*, Vol. 45, 1997, pp. 1681–1688.

[20]   McGrath, W. R., et al., "Silicon Micromachined Waveguides for Millimeter-Wave and Submillimeter-Wave Frequencies," *IEEE Microwave and Guided Wave Letts.*, Vol. 3, 1993, pp. 61–63.

[21]   Lorenz, H., et al., "SU-8: A Low-Cost Negative Resist for MEMS," *J. Micromech. Microeng.*, Vol. 7, 1997, pp. 121–124.

[22]   Osorio, R., et al., "Micromachined Strip Line with SU-8 as the Dielectric," *11th GAAS Symposium*, Munich, Germany, 2003, October 6–10, pp. 179–182.

[23]   Ermolov, V., et al., "Microreplicated RF Toroidal Inductor," *IEEE Trans. on Microwave Theory Tech.*, Vol. 52, No. 1, January 2004, pp. 29–37.

[24]   De Los Santos, H. J., "On the Ultimate Limits of IC Inductors—An RF MEMS Perspective," IEEE *Electronic Components and Technology Conference*, 2002, San Diego, CA: pp. 1027–1031.

[25]   Chang, J. Y.-C., A. A. Abidi, and M. Gaitan, "Large Suspended Inductors on Silicon and Their Use in a 2 mm CMOS RF Amplifier," *IEEE Electron Device Letts.*, Vol. 14, 1993, pp. 246–248.

[26]   Lopez-Villegas, J. M., et al., "Improvement of the Quality Factor of RF Integrated Inductors by Layout Optimization," *IEEE Trans. Microwave Theory Tech.*, Vol. 48, No. 1, 2000, pp. 76–83.

[27]   Jiang, H., Y. Wang, J.-L. A Yeh., and N. C. Tien, "On-Chip Spiral Inductors Suspended over Deep Copper-Lined Cavities," *IEEE Trans. Microwave Theory and Techniques*, Vol. 48, No. 12, December 2000, pp. 2415–2423.

[28]   Yoon, J.-B, et al., "Surface Micromachined Solenoid On-Si and On-Glass Inductors for RF Applications," *IEEE Electron Device Letters*, Vol. 20, 1999, p. 487.

[29]   Liu, W. Y., et al., " Toroidal Inductors for Radio-Frequency Integrated Circuits," *IEEE Trans. Microwave Theory Tech.*, Vol. 52, No. 2, February 2004, pp. 646–654.

[30]   Dahlmann, G. W., et al., "MEMS High Q Microwave Inductors Using Solder Surface Tension Self-Assembly," *2001 IEEE IMS Digest of Papers.* May 20–25, pp. 329–332.

[31]   Lubecke, V. M., et al., "Self-Assembling MEMS Variable and Fixed RF Inductors," *IEEE Trans. Microwave Theory Tech.*, Vol. 49, No. 11, November 2001, pp. 2093–2098.

[32]   Chen, J., et al.,"Design and Modeling of a Micromachined High-Q Tunable Capacitor With Large Tuning Range and a Vertical Planar Spiral Inductor," *IEEE Trans. Electron Dev.*, Vol. 50, No. 3, March 2003, pp. 730–739.

[33]   Sun, Y., et al., "Suspended Membrane Inductors and Capacitors for Application in Silicon MMICs," *IEEE Microwave and MillimeterWwave Monolithic Circuits Symposium Digest of Papers*, 1996, pp. 99–102.

[34]   Abidi, A. A., "Low-Power Radio-Frequency ICs for Portable Communications," *Proc. IEEE*, 1995, pp. 544–569.

[35]   Dec, A., and K. Suyama, "Micromachined Varactor with Wide Tuning Range," *Electronic Letts.*, Vol. 33, 1997, pp. 922–924.

[36]   Young, D. J., and B. E. Boser, "A micromachined Variable Capacitor for Monolithic Low-Noise VCOs," *Solid-State Sensor and Actuator Workshop*, Hilton Head, SC, June 2–6, 1996, pp. 86–89.

[37]   Hung, E. S., and S. D. Senturia, "Tunable Capacitors with Programmable Capacitance-Voltage Characteristic," *Solid-State Sensor and Actuator Workshop*, Hilton Head, SC, June 7–11, 1998.

[38]   Ionis, G. V., A. Dec, and K. Suyama, "Differential Multifinger MEMS Tunable Capacitors for RF Integrated Circuits," *2002 IEEE MTT-S Int. Microwave Symposium Digest*, 2002, pp. 345–348.

[39]   Feng, Z., et al., "MEM-Based Variable Capacitor for Millimeter-Wave Applications," *Tech. Digest Solid-State Sensors and Actuators Workshop*, Hilton Head Island, SC, 2000, pp. 255–258.

[40]   Oz, A., and G. K. Fedder, "RF CMOS-MEMS Capacitor Having Large Tuning Range," *Transducers'03, The 12th International Conference on Solid State Sensors, Actuators and Microsystems*, Boston, MA, June 8–12, 2003, pp. 851–854.

[41]   Nieminen, H., et al., "Microelectromechanical Capacitors for RF Applications," *J. Micromech. Microeng.*, Vol. 12, 2002, pp. 177–186.

[42]   Tsang, T. K., et al., "Wide Tuning Range RF-MEMS Varactors Fabricated Using the PolyMUMPs Foundry," *Microwave Journal*, August 2003.

[43]   Fischer, G., W. Eckl, and G. Kaminski, "RF-MEMS and SiC/GaN as Enabling Technologies for a Reconfigurable Multi-Band/Multi-Standard Radio," *Bell Labs Technical Journal*, Vol. 7, No. 3, 2003, pp. 169–189.

[44]   Zine-El-Abidine, I., M. Okoniewski, and J. G. McRory, "A New Class of Tunable RF MEMS Inductors," *IEEE Proceedings of the International Conference on MEMS, NANO and Smart Systems* (ICMENS'03). July 20–23. Banff, Alberta, Canada. pp. 114–117.

[45]   Tassetti, C.-M., G. Lissorgues, and J.-P. Gilles, "Effects of a Loop Array Layer on a Microinductor for Future RF MEMS Components," *33rd European Microwave Conference*, Munich, Germany, 2003, pp. 29–32.

[46]   Yokoyama, Y., et al., "On-Chip Variable Inductor Using Microelectromechanical Systems Technology," *Jpn. J. Appl. Phys.*, Vol. 42, 2003, pp. 2190–2192.

6

MEMS-Based Microwave Circuits and Systems

6.1 Introduction

As consumer demand for ever more powerful wireless products and services is expected to continue for the foreseeable future [1], research on device and systems technologies with the potential to reconcile conflicting expectations, such as portability and low cost, together with unprecedented levels of functionality, flexibility and sophistication, are being vigorously pursued [1, 2]. While the functionality in these systems stems from their ability to provide multiple simultaneous functions, such as telephone, fax communications, and computing, the flexibility and the sophistication derive from the powerful features enabled by digital signal processing techniques [1]. The specific products in question include lightweight voice, data terminals, small wireless data terminals, and notebook hand-held computers. The related services, in turn, include access to network wireline services via wireless terminal devices, as well as wireless communications with quality comparable to that of wireline networks. The unifying theme permeating these consumer products and services is obtaining *portability and low-cost*, which, in turn, is enabled by reduced power consumption. In particular, the overarching theme in recent years is the convergence of multiple functions onto a single portable wireless appliance [3].

The above situation is not much different in the area of satellite communications, where, at launch, costs nearing the tens of thousands of dollars per kilogram [4] of *payload mass*, the drive to reduce spacecraft mass in view of increasing customer demands for onboard features like multiuser, multidata rate, and multilocation satellite links [5, 6], is at a premium. Unlike current mobile- and

173

cellular-based ground systems, whose frequency of operation lies in the 900-MHz to 6-GHz range [2], satellite-based communications systems operate at microwave- and millimeter-wave frequencies, in the 4- to 100-GHz range. This means that candidate technologies exhibiting just low-power consumption are not totally useful, but rather, must be accompanied by high-frequency capability.

Furthermore, the ever-increasing proliferation [7] of *wireless standards* has impacted the number of base station products and designs that infrastructure providers must deploy and maintain [7]. Wireless standards are sets of specifications (i.e., frequency of transmission and reception), which dictate the hardware and software implementation of wireless appliances. Those wireless appliances operating under the same standard will communicate with each other, but not with appliances operating under a different standard.

As will be apparent from the subsequent presentation, there are two aspects to enabling portability, low cast, and reduced mass in wireless systems. These are the development of very low loss passive components, on the one hand, and of very efficient active components, on the other hand. By passive components, we refer loosely to devices and circuits, like transmission lines, whose operation does not require the application of standby dc power (e.g, a battery), and that are characterized by the loss they introduce on a signal traversing them. By active devices and circuits, on the other hand, we refer to those that do require the application of standby dc power for their operation (e.g., amplifiers and digital circuits).

Our presentation in previous chapters should have made it clear that MEMS technology has a tremendous potential to greatly improve the performance of *both* passive and active circuits. In naming a few examples, we might mention that micromachining of the substrate beneath inductors and transmission lines improves their quality factor (exhibiting lower insertion loss), and extends their frequency of operation; that MEM techniques can produce on-chip variable capacitors with potentially higher Q than semiconductor-based varactors; and that MEM switches have the potential for very low insertion loss, high bandwidths, and zero standby dc power consumption.

In this chapter, we shall conclude our introduction to MEMS by first reviewing the fundamentals of these systems, and then delineating the role that MEMS-based circuits and systems can play to enable the paradigms of portability and low cost, of sophisticated satellite communications, and of versatile base stations. Since the readership of this book is expected to come from a wide range of technical backgrounds (e.g., from materials, process and device scientists and engineers, who may not be familiar at all with RF and microwave electronics, to microwave engineers proper), our presentation follows a semi-intuitive approach, making use of mathematics as the common language. Readers with a background in microwave electronics may wish to skip the sections on "fundamentals."

6.2 Wireless Communications Systems

6.2.1 Fundamentals of Wireless Communications

The fundamental problem of wireless communications consists of transferring information between a source and a destination, as shown in Figure 6.1. If the signal representing the information to be transmitted is in analog form (i.e., exists at all times), a straightforward way to accomplish the transmission would be to feed the signal to an antenna (at the source) which will convert it into an electromagnetic (EM) wave and radiate it into space. The EM wave, after propagating through space, will reach the desired destination. If the signal at the source represents, for example, music or voice, which have a maximum frequency content of about 20 kHz, this would require that, for reasonable efficiency, the transmitting antenna required would have dimensions of half-wavelength, given by (6.1).

$$\frac{\lambda}{2} = \frac{c}{2f} = \frac{3 \times 10^8 \text{ m/s}}{2\left(20 \times 10^3 \text{ m/s}\right)} = 7,000\text{m} \approx 4.35 \text{ mi} \qquad (6.1)$$

This antenna size is both impractical and expensive. Now, examination of the above equation reveals that we could bring the antenna down to a reasonable size if the signal consisted of a higher frequency. Unfortunately, music consists of low frequencies, or we might actually be interested in transmitting low frequency signals (e.g., seismic information). The question arises then as to how to resolve this situation. The most widely adopted solution to this dilemma is to use a high-frequency signal to carry the desired information. The process by which the desired information is impressed upon the high-frequency signal is called *modulation*. At the destination, the inverse process (i.e., *demodulation*) is performed on the received signal in order to extract the desired information from the high-frequency signal. The set of frequencies that comprise the desired information to be transmitted, $m(t)$, in its original form is referred to as the *baseband*. On the other hand, the continuous wave that carries the baseband signal is referred to as the *carrier*. Modulation of the carrier is effected when the

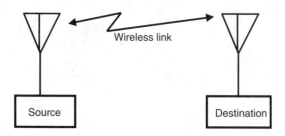

Figure 6.1 Wireless communications paradigm.

amplitude, the frequency, or the phase of the carrier is made to vary in a manner dictated by the baseband signal. Accordingly, we can have four fundamental types of modulation schemes, namely, amplitude modulation (AM), frequency modulation (FM), phase modulation (PM), and digital modulation.

6.2.1.1   Ubiquitous Wireless Connectivity

The ability to achieve global communications via wireless connectivity is predicated upon providing means to route an information signal from an origin to a destination, regardless of the distance between them. Since the transmission power available to battery-powered portable wireless appliances is rather limited, the signals they transmit may be attenuated upon propagating several hundred meters, thus becoming essentially undetectable. It is necessary, therefore, to provide an infrastructure (i.e., a network of regeneration platforms to amplify and retransmit these signals). When these relaying platforms are located on the Earth, they are called *base stations*. This infrastructure is acquired and maintained by, for example, a cell phone service provider. Similarly, when these relaying stations are deployed in outer space, they are called *satellites*. Figure 6.2 shows a sketch that captures the ubiquitous wireless connectivity paradigm.

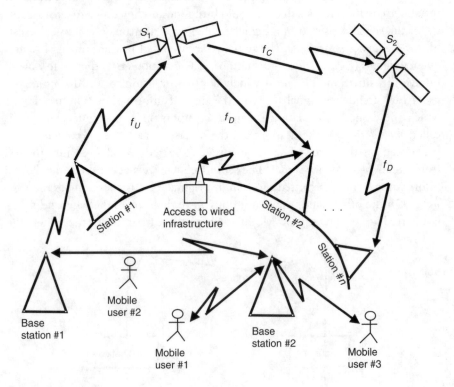

Figure 6.2 Ubiquitous wireless connectivity paradigm.

Next, we discuss the fundamentals of mobile platforms, base stations, and satellites.

6.2.1.2 Mobile Platforms

Whatever the modulation scheme utilized, wireless communications systems are functionally partitioned into two parts: the front-end, which processes the modulated high-frequency carrier, and the baseband or back-end, which processes the low-frequency baseband signal. Therefore, a typical receiver architecture, as shown in Figure 6.3, consists of an antenna to pick up EM energy around the carrier frequency, f_{RF}; a high-frequency *bandpass* filter to select the carrier signal from among all signals present at the antenna; an amplifier to raise the power level of the selected carrier; an oscillator and mixer (a nonlinear frequency translation circuit) to translate the carrier frequency to a lower intermediate frequency, f_{IF}, at which the demodulation operation might be easier to perform; and a second lower-frequency filter to remove unwanted frequencies produced during the mixing operation.

The increasing number of wireless standards, and the desire to implement multiband and multistandard wireless appliances embodying the convergence of these, has elicited the need for flexible transceiver architectures. In this context, Kaiser [8] has proposed a variety of reconfigurable architectures in which RF MEMS plays an enabling role, as shown in Figure 6.4. From an examination of these transceiver architectures, one can surmise that RF MEMS may play an enabling role in the following functions: transmit/receive switch, filtering, matching networks, and reconfigurability. Table 6.1 indicates specific RF MEMS component opportunities to enable these functions.

6.2.1.3 Base Stations

Base stations extend the geographical coverage of wireless appliances (see Figure 6.2), by providing a communications link between users separated by distances beyond the reach of their battery-powered equipment. Thus, they consist of relatively powerful receiver/transmitter units, through which wireless traffic is amplified and rerouted. In this context, the architecture of a base station

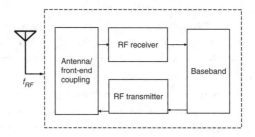

Figure 6.3 Conceptual receiver/transmitter (transceiver) architecture.

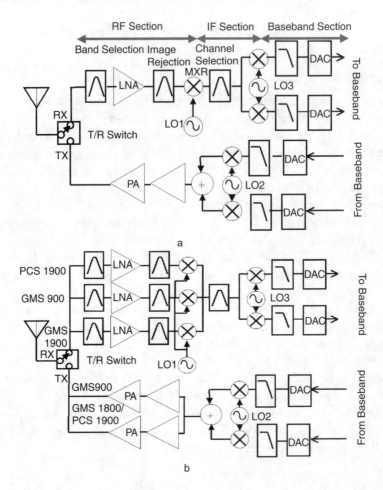

Figure 6-4 Proposed transceiver architectures for flexible wireless appliances: (a) TDMA transceiver architecture, (b) multiband transceiver architecture, (c) multistandard transceiver architecture, and (d) software-defined radio. (e) Zero-IF transceiver. (*After.* [8].)

transceiver is similar to that of a mobile platform, as shown in Figure 6.3. As part of the wireless infrastructure, however, base stations tend to be deployed in a permanent fashion, and must be capable of handling and accommodating the rapidly increasing number of wireless standards. This poses an expensive proposition for base station providers [7] who, in order to serve different markets, characterized in turn by different frequency bands and different standards, find themselves in the predicament of having to maintain a broad selection of products in stock, and are unable to satisfy market needs due to the long duration of new product design cycles.

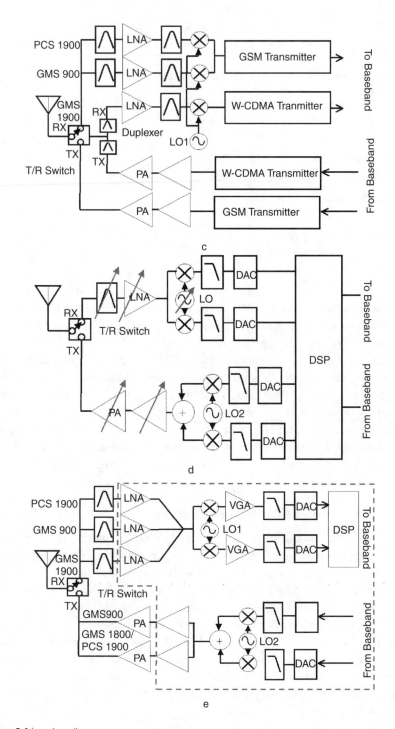

Figure 6.4 (continued)

Table 6.1
Opportunities for RF MEMS Components in Mobile Platforms

System Function	RF MEMS Components			
	Switch	Inductor/Variometer	Capacitor/Varactor	Resonator
Antenna-matching network	X	X	X	
T/R switch	X			
Duplexer/diplexer (filters)				X
Smart antennas	X			
(phase shifters/power combining)				
Band reconfiguration	X	X	X	X
Transceivers (LNAs, VCOs, PAs)	X	X	X	X

To address these issues, the concept of reconfigurable multiband and multistandard architecture for base stations has been introduced by Fischer, Eckl, and Kaminski [7]. This concept is predicated upon the utilization of RF MEMS components to enable superior reconfigurable functions, in particular: (1) reconfigurable receive and transmit filters, (2) tunable preselector and postselector analog filters, (3) reconfigurable (switchable) multiband frequency synthesizer, (4) reconfigurable baseband filters to appropriately adopt the bandwidth accompanying different standards, (5) reconfigurable impedance matching networks to optimize matching upon switching bands, and (6) multiband power amplifiers with optimized (narrow) bandwidth and power efficiency. Table 6.2 describes opportunities for RF MEMS components in base stations.

6.2.2   Fundamentals of Satellite Communications

When the source and destination are a large distance apart so that, due to the Earth's curvature, straight line-of-sight EM wave propagation is not possible, the communications link is provided by a satellite, as shown in Figure 6.3. The transmitting station launches a carrier signal of a certain *uplink* frequency, f_U, into space in the direction of the satellite. The receiving station, on the other hand, is equipped to receive a carrier signal of another *downlink* frequency, f_D, from the direction where the satellite floats in space. In addition, the satellite might also be equipped to communicate with other satellites through *a crosslink* frequency, f_C. In general, the more uplinks, crosslinks, and downlink frequencies a satellite can maintain, the greater is its capacity and usefulness; unfortunately, its required physical size (mass) and power consumption is also greater. Since mass is the primary driver of satellite costs [5, 6], it becomes tightly coupled to the satellite's performance, ultimately playing a limiting role in it.

Table 6.2
Opportunities for RF MEMS Components in Base Stations

System Function	RF MEMS Components			
	Switch	Inductor/ Variometer	Capacitor/ Varactor	Resonator
Multiband/broadband antenna	X			
Reconfigurable receive and transmit filter	X			X
Tunable preselector and postselector		X	X	
Reconfigurable multiband synthesizer	X			
Reconfigurable baseband filter		X	X	X
Reconfigurable impedance matching		X	X	
Multiband power amplifiers	X	X	X	

The communications satellite could be viewed as composed of two main parts [4], namely, a *platform* and a *payload*. The platform includes the following subsystems: (1) the physical structure; (2) the electric power supply; (3) temperature control; (4) attitude and orbit control; (5) propulsion equipment; and (6) telemetry, tracking and command equipment. The payload, on the other hand, includes: (1) the receiving antenna, (2) the transmitting antenna, and (3) all electronic equipment supporting transmission of carriers.

An examination of the mass and power distribution in a conventional geosynchronous satellite [5] reveals that the communications payload accounts for roughly one quarter of the spacecraft's dry mass, and that it consumes the most power of any single subsystem. Furthermore, within the payload, the transmitters and antennas account for the bulk of the mass and the power consumption. These fractions are bound to increase even more as emerging satellite architectures incorporate increasing capacity and flexibility [4–6]. In particular, the capacity and flexibility possessed by a satellite are strongly dependent on the agility of its antenna, with the technology of choice being the so-called phased-array antenna.

6.2.2.1 Phased-Array Antennas

The ability of satellites to provide *dynamic* links to multiple users in real time is enabled by the phased-array antenna. In simple terms, a phased-array antenna consists of a set of phase shifters, or true time delay units, that control the amplitude and phase of the excitation to an array of antenna elements, in order to set the beam phase front in a desired direction, as shown in Figure 6.5. In a typical Ku-band 5-bit state-of-the-art phasor downlink module, each channel, as shown

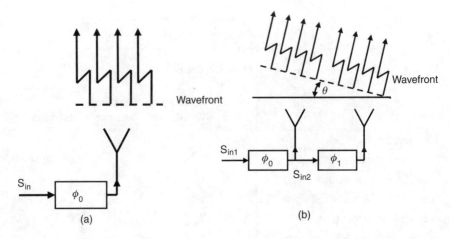

Figure 6.5 Phased-array antenna concept: (a) Single-antenna radiation, and (b) phased-antenna radiation.

in Figure 6.6 could consist of a phase shifter, an attenuator, and a solid state power amplifier (SSPA) , implemented in MMIC technology, driving an array of antenna elements. An examination of the typical measured performance of the individual blocks [6] making up a channel reveals a number of technology-related limitations, the most prominent of which has to do with the huge insertion loss (e.g., 10.5 dB @14.5 GHz) introduced by a state-of-the-art FET-based 5-bit phase shifter MMIC chip. The phase shifter, despite possessing wideband performance and small size, dominates the loss of the chain, thus placing an undue burden on the SSPA by demanding a higher gain and power consumption. These, in turn, drive the unit's power supply capability, heat sinking, and weight requirements. When one considers that full-scale phased-array antennas contain thousands of channels, it becomes obvious that such losses as exhibited by FET-based phase shifters are prohibitive. The fundamental reason for the high insertion loss associated with FET-based phase shifters lies in the inevitable device channel resistances: both the "open" channel resistance for the case of the low-impedance state, and the residual series resistance in the pinch-off channel,

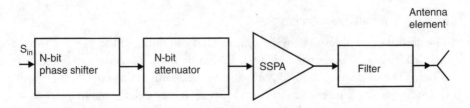

Figure 6.6 Typical phased-array antenna channel.

for the high-impedance case [9]. This insertion loss problem of conventional phase shifters presents itself as an opportunity for MEMS-based phase shifters. Indeed, MEM switches (see Chapter 4) have the potential for ultralow insertion loss to minimize the use of SSPAs; are capable of broadband operation, to achieve versatility for diverse and simultaneous tasks; possess high electrical isolation, to minimize crosstalk effects; possess ultralight weight (mass), to effect a lower cost per payload; and are inexpensive to manufacture. Furthermore, a comparison of the performance of MEM and PIN-diode switches shows that MEM switches are far superior, both in terms of insertion loss and isolation, as well as bandwidth [9]. Therefore, it is clear that MEM switches are posed to become a key element in the phased-array antennas of the not-too-distant future.

6.3 MEMS-Based RF and Microwave Circuits

The fundamental functions of RF and microwave circuits include detecting, producing, or processing high-frequency electronic signals, where every signal is characterized by the frequency (or set of frequencies) it contains, called its *frequency spectrum*, and by its amplitude or power level. By signal processing, we refer to the acts of effecting changes on the frequency contents of the original signal, of changing its time location (phase) relative to other signal(s), or of scaling its amplitude. In the following sections, we review the fundamentals of some of the key RF and microwave circuits employed to realize these functions. In particular, we present the fundamentals of phase shifters, filters, oscillators, and mixers.

6.3.1 Phase Shifters

As seen in the previous section, phase shifters are key to the realization of phased-array antennas [10], as well as many other radar and communications systems [11]. A phase shifter is a circuit that, ideally, produces a replica of a signal applied at its input, but with a modified phase, as shown in Figure 6.7(a). Since a fundamental unit of phase shift can be produced by a few basic building blocks (e.g., RC, LC, and transmission line networks [11]), any value of phase shift can be produced on a signal by combining that which is produced by these elemental units, as shown in Figure 6.7(b). A switched-line phase shifter implementation is shown in Figure 6.7(c). The performance of a phase shifter is characterized by its insertion loss, bandwidth, power dissipation, power handling capability, and insertion phase. A recent thorough discussion of MEMS-based phase shifters, together with prototype design and measured performance data, is given in [12, 13].

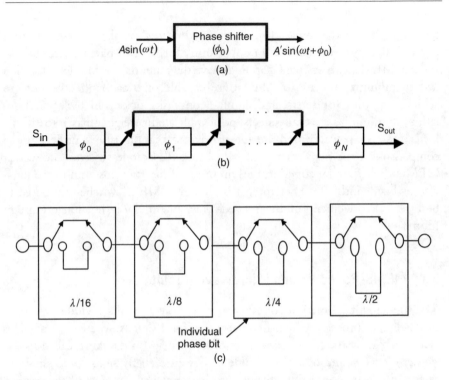

Figure 6.7 Phase shifter: (a) System-level view, (b) building-block view, and (c) switched-line implementation.

6.3.2 Resonators

6.3.2.1 Resonator Fundamentals

As is well known, at frequencies up to around 100 MHz, LC resonators are realized as series or parallel connections of inductors and capacitors. At higher frequencies, however, the physical dimensions of inductors and capacitors utilized in these LC resonators become of the order of the signal wavelength, $\lambda = c/f$. When this occurs, the propagation delay of the signal as its traverses these elements is no longer negligible with respect to the signal period. That is, currents and voltages in the inductor and capacitors no longer may be assumed to be constant throughout their extent, and Maxwell's equations for the *spatial* propagation of electromagnetic waves must be solved to obtain the voltage and current distributions. This is the so-called *distributed-circuit* regime. In addition to becoming distributed (wave) circuits, the coincidence of signal wavelength with component physical dimension triggers radiation effects. Thus the circuits become lossy, and stop behaving as energy storage elements. Finally, conduction losses due to the onset of the skin effect in the conductors become noticeable. To avoid energy loss due to radiation and attenuation due to the skin effect, a structure that would, on the one hand, provide shielding against radiation and, on the other hand, minimize the skin effect resistance, was arrived at [14, 15], namely, a waveguide-based rectangular cavity. The cavity resonator is a metallic box that has the ability to confine (store) electromagnetic energy very efficiently [i.e., with rather small losses (high Q)]. Unfortunately, waveguides tend to be bulky, to be costly, and, as discussed in Chapter 4, to be incompatible with conventional planar fabrication techniques.

6.3.2.2 Micromechanical Resonator Circuits

As discussed in Chapters 3 and 4, MEMS technology has been exploited to realize micromechanical resonators in the form of one-port or two-port vertically vibrating cantilever beams, or as laterally vibrating comb-driven folded-beam structures, and more recently, as capacitively transduced lateral-vibrating clamped–clamped beams [16]. Of these structures, the laterally vibrating ones are generally preferred, because their linear force-capacitance relationship enables higher dynamic range, and because very high quality factors (e.g., on the order of 80,000), are achievable in vacuum [16]. These features, together with IC-fabrication compatibility and the ability for on-chip frequency and Q trimming [17], contribute to their large potential as enablers of single-chip wireless communications systems; that is, they make unnecessary the use of off-chip bulkier LC components.

In order to utilize these resonators within a circuit, it is necessary to couple to their vibration. Coupling to a micromechanical resonator can be accomplished in two ways, namely, via mechanical coupling, or via electrical coupling. Mechanical coupling is realized by attaching a structure (i.e., a *spring*), to the resonator that can propagate or extract some of its motional mechanical energy without greatly disturbing or *loading* it, as shown in Figure 6.8(a, b). Electrical coupling, on the other hand, is accomplished via, for example, a capacitive transducer, such as the interdigitated comb-drive capacitor system. In the case of electrical coupling, since the resonator's sense electrode output signal is a current, satisfying the requirement of negligible resonator loading requires that it be followed by a *buffer* transimpedance (current in/voltage out) amplifier, as shown in Figure 6.8(c). The effect of loading a resonator is to lower the overall Q of the resulting circuit. This is expressed in terms of the resonator's *loaded Q*, Q_L, given by

$$\frac{1}{Q_L} = \frac{1}{Q_e} + \frac{1}{Q} \qquad (6.2)$$

where $Q = \omega_0 L / R$ and $Q_e = \omega_0 L / R_L$ for series resonant circuits; $Q = \omega_0 RC$ and $Q_e = R_L / \omega_0 L$ for parallel resonant circuits; and R_L is the external load resistor which

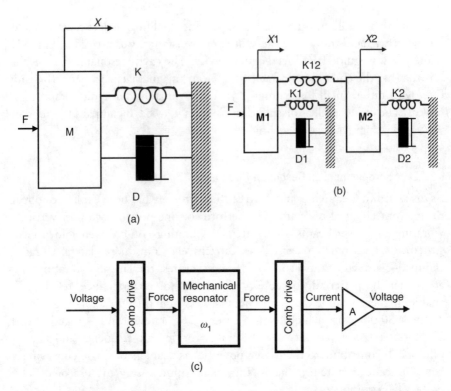

Figure 6.8 Mechanical model of resonator: (a) Single resonator, and (b) mechanically-coupled resonator. (*After:* [18].) (c) Electrically-coupled resonator.

adds to the internal resonator resistance R. This added external loss manifests as an insertion loss (I.L.) of the resonator at resonance, given by [19]:

$$I.L.(dB) = 20\log_{10}\left(1 + \frac{Q_L}{Q}\right) \tag{6.3}$$

In the next section, we will deal with the micromechanical resonator as a building block for filters.

6.3.3 Filters

6.3.3.1 Filter Fundamentals

Before a signal can be processed, it is necessary to select it from among all the additional signals that, possibly due to interference effects, may be simultaneously present. Therefore, perhaps the most fundamental RF and microwave circuit function is filtering. If one is interested in selecting a signal containing all the frequencies below a certain maximum or cutoff frequency, ω_c, then one needs a *lowpass* filter, as shown in Figure 6.9(a).

If, on the other hand, one is interested in selecting a signal whose frequency components lie above a minimum frequency, then one needs a *highpass* filter, as shown in Figure 6.9(b). Finally, if one is only interested in selecting a limited set of frequencies (e.g., those in the band $\omega_{cU} - \omega_{cL}$, around a center frequency ω_c, then one needs a *bandpass* filter, as shown in Figure 6.9(c). Another filter that is often used is the bandstop filter. In this case, one is interested in selecting all frequencies except those in the band $\omega_{cU} - \omega_{cL}$. In addition to its frequency response, a function of interest in many filter applications is the *phase* response. In particular, a filter with a *linear phase versus frequency* response

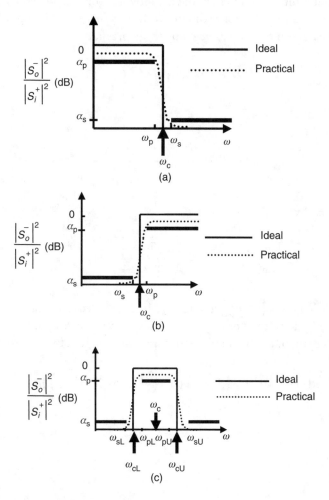

Figure 6.9 Ideal filter responses: (a) Lowpass, (b) highpass, and (c) bandpass.

exhibits a *constant delay* at all frequencies within its passband. This is a desirable feature because it means that all inband frequency components will experience the same delay, and thus signal distortion due to frequency dispersion will be averted.

Unfortunately, it is physically impossible to realize filters with the ideally sharp transmission characteristics shown in Figure 6.9. Therefore, in order to construct a practical filter, it is necessary to specify the degree to which its characteristics must approximate the ideal ones. Towards that end, there exist both mathematical expressions, which relate the desired filter response to its parameters and number of elements (filter *degree*), thus helping to determine the details of the filter which will result in the desired response, as well as extensive tables on filter parameters for obtaining specific responses [20]. Two particularly simple and popular filter responses are the Butterworth (or maximally flat), and the Chebyshev (or equiripple) approximations. In the Butterworth approximation, the filter attenuation response is given by the expression

$$\alpha(\omega) = 10 \log_{10}\left[1 + C^2 \omega^{2N}\right] \tag{6.4}$$

where C may be chosen as

$$C = \frac{\sqrt{10^{\alpha_s/10} - 1}}{\omega_s^N} \tag{6.5}$$

and N is given by

$$N \geq \frac{\log_{10} k_1}{\log_{10} k}, \quad k_1 = \sqrt{\frac{10^{\alpha_s/10} - 1}{10^{\alpha_s/10} - 1}}, \quad k = \frac{\omega_p}{\omega_s} \tag{6.6}$$

where k is called the *selectivity* parameter [21]. It is called maximally flat because the first $N - 1$ derivatives of the response are zero at the origin. In the Chebyshev approximation, the filter attenuation response is given by the expression

$$\alpha(\omega) = 10 \log_{10}\left[1 + k_p^2 \cos^2\left(N \cdot \cos^{-1}(\omega/\omega_p)\right)\right] \tag{6.7}$$

with

$$k_p = \sqrt{10^{\alpha_p/10} - 1} \tag{6.8}$$

and N given by

$$N \geq \frac{\cosh^{-1}\left(1/k_1\right)}{\cosh^{-1}\left(1/k\right)} \qquad (6.9)$$

For a given filter degree N, the Chebyshev approximation (called equirip-ple because, within the passband, the response fluctuates at a constant amplitude around the mean value), gives a sharper transition region than the Butterworth approximation.

Now, since all frequencies appear the same at a given instant of time, discriminating among signals of different frequencies (i.e., performing frequency filtering), requires that the frequencies of the signal under observation be compared with the reference frequency and frequencies of interest over a period of time. In other words, it is necessary to employ a circuit with memory [22]. The most fundamental "frequency storage" device, realizing this memory element, is the resonator circuit. The frequency is stored in the rate at which energy present in the resonator is transformed back and forth between two physical forms, such as kinetic and potential, as in the case of the cantilever beam, and electric and magnetic, as in the case of the familiar LC- and microwave cavity-resonators, depending on the physical implementation of the circuit.

Regardless of the form of the resonator utilized, a bandpass filter can be conceptually represented as a chain of resonators, depicted in Figure 6.10(a), in which the first and last resonators are coupled to the input and output, respectively, via corresponding coupling coefficients K_I and K_o. These coefficients give an indication of the efficiency with which incoming signals can enter the input "port," and exit the output "port." The inner resonators are coupled to their nearest neighbors via coupling coefficients K_{ij}. To explain the operation of the

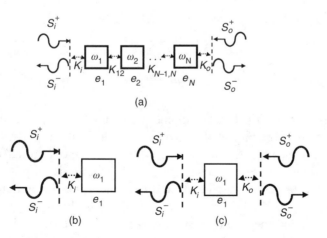

(a)

(b)   (c)

Figure 6.10 (a) Conceptual representation of resonator-coupled bandpass filter, (b) single (load) resonator, and (c) transmission resonator.

filter, consider first the case of the single resonator, as shown in Figure 6.10(b). Suppose the resonator is somehow energized. Then, by its very nature, its energy will oscillate between two physically different forms at a rate given by its resonance frequency, in this case, ω_1. An ideal resonator would continue oscillating forever without experiencing any decay in its energy. However, in reality, intrinsic energy dissipation mechanisms cause the stored energy to decrease at a rate given by a characteristic time constant, $1/\tau_0$, which embodies the magnitude of the losses. These losses are characterized by the resonator's quality factor, which is given by $Q = 2/\omega_1\tau_0$ [23]. In addition, if the resonator is not isolated (i.e., it can exchange energy with its surroundings), in particular with the input "port," then two other mechanisms can change its energy. First, energy can "leak" at a rate given by a time constant $1/\tau_{ei}$, and second, an incoming signal of amplitude S_i^+ and carrying power $|S_i^+|^2$, will *add* an amount $K_I \cdot S_i^+$ to the rate of increase of resonator energy. In summary, then, the rate of increase in resonator energy obeys the following balance equation [23]:

$$\frac{de_1}{dt} = j\omega_1 e_1 - \frac{e_1}{\tau_0} - \frac{e_1}{\tau_{ei}} + K_i S_i^+ \qquad (6.10)$$

If it is assumed that the incoming signal varies with time, $S_i^+ \propto (j\omega t)$, then the resonator energy amplitude will have a frequency response given by:

$$e_1 = \frac{K_i S_i^+}{j(\omega - \omega_1) + \dfrac{1}{\tau_0} + \dfrac{1}{\tau_{ei}}} \qquad (6.11)$$

and its power will vary as

$$|e_1|^2 = \frac{K_i^2 \tau^2}{(\omega - \omega_1)^2 \tau^2 + 1} \cdot |S_i^+|^2 \qquad (6.12)$$

where $1/\tau = 1/\tau_0 + 1/\tau_{ei}$. This equation expresses the fact that resonator power is sensitive to the frequency of the incoming signal. In particular, when the frequency of the incoming signal coincides with its resonance ("stored") frequency, the power transfer from the incoming signal into the resonator peaks.

Next, consider the case in which the resonator *also* has an output port, as shown in Figure 6.10(c). In this case, the coupling to the output port is characterized by K_o, and the energy "leakage" via this port is characterized by a time constant $1/\tau_{eo}$. The rate of *increase* of resonator energy now obeys the following general balance equation [23]:

$$\frac{de_1}{dt} = j\omega_1 e_1 - \frac{e_1}{\tau_0} - \frac{e_1}{\tau_{ei}} - \frac{e_1}{\tau_{eo}} + K_i S_i^+ + K_o S_o^+ \tag{6.13}$$

If we assume that no signal impinges unto the resonator at its output port (i.e., $S_o^+ = 0$), then the resonator power varies also as

$$\left|e_1\right|^2 = \frac{K_i^2 \tau^2}{\left(\omega - \omega_1\right)^2 \tau^2 + 1} \cdot \left|S_i^+\right|^2 \tag{6.14}$$

but now, because of the coupling K_o, an amount of power $\left|S_o^-\right|^2 = K_o^2 \cdot \left|e_1\right|^2$ can be transferred to the output port. Combining this equation with (6.14), we obtain the power transmitted to the output port *from* the input port as

$$\left|S_o^-\right|^2 = \frac{K_i^2 K_o^2 \tau^2}{\left(\omega - \omega_1\right)^2 \tau^2 + 1} \cdot \left|S_i^+\right|^2 \tag{6.15}$$

This expression tells us that the output signal power is a frequency *selective* function of the input signal frequency (i.e., *the system is a filter*).

The performance (i.e., selectivity of the single resonator filter), which has degree $N = 2$, as shown by the degree of the denominator polynomial in (6.15), is rather limited and may not be adequate. Greater selectivity is obtained by increasing N, which entails increasing the number of coupled resonators, as shown in Figure 6.10(a). To arrive at the desired filter, use is made of cookbook-type references (e.g., the "Tables of k and q Values" [20]), which contain *frequency-normalized* data on the filter attenuation characteristics, and on the corresponding coupling coefficients needed to produce a given approximation. The selection of these parameters is clearly illustrated in [20], p. 309 for the Butterworth response, and will not be repeated here.

6.3.3.2 Surface-Micromachined MEM-Resonator Filters

Mechanical-resonator-based filters are not new. Indeed, vestiges of mechanical filter concepts appeared as early as 1926, when Maxfield and Harrison, in order to improve the frequency response of a phonograph, applied electrical network theory and electromechanical analogies to treat the needle and the needle arm as a network. In this context, a capacitor was utilized to represent the needle compliance, and an inductor to represent the needle arm mass [21, 24]. The motivation for using mechanical resonators is traced [24] to their high Q factors (e.g. between 10,000 and 25,000), and to their good stability, which allowed the production of very narrowband filters (i.e., 0.1% at 455 kHz). Unfortunately, the excellent performance of mechanical filters came at a price: they carry a higher manufacturing cost and tend to be bulky. These characteristics, together with the conception of new filtering solutions for integration with ICs [25], relegated them to certain niche applications. The subject of mechanical filters, and in particular microelectromechanical filters, has received renewed attention [26–30] because of the possibility of bringing the excellent filtering properties that mechanical resonators provide into the realm of planar ICs. Indeed, Nguyen [26–28, 30] has been involved extensively in pioneering the development of MEM filters.

The design of resonator-coupled filters hinges upon obtaining coupling coefficients that will enable a given response to be met. Families of normalized filter responses are cataloged in the "Tables of k and q Values" [20]. Once the filter degree that will meet the desired selectivity is obtained, the point of entry into these tables is the normalized resonator quality factor, q_0, defined by

$$q_0 = \frac{\Delta f}{f_c} Q_0 \qquad (6.16)$$

where Δf is the bandwidth, f_c is the center frequency, and Q_0 is the intrinsic resonator quality factor of the resonators, assumed uniform. The normalized coupling coefficients and quality factors, $k_{i,j}$ and q_i, respectively, can subsequently be read off the tables. From these, the denormalized coupling coefficients and quality factors are obtained according to

$$K_{i,j} = k_{i,j} \frac{\Delta f}{f_c} \qquad (6.17)$$

and

$$Q_i = q_i \frac{f_c}{\Delta f} \qquad (6.18)$$

To particularize the coupling coefficients to the case of MEM resonators, the MEM filter is usually simulated using an LCR electrical circuit based on the equivalent circuit models of the resonators. Once L_x, C_x, and R_x are obtained for each resonator, the *electrical* design of the filter is done, and then its *mechanical* design may begin. The first step in this direction is to obtain the effective mass of the resonators, $M_{i,res}$. For a comb-driven resonator, this is related to the inductance, L_x, by

$$M_{i,res} = \eta_i^2 L_{i,x} \qquad (6.19)$$

where $\eta_i = V_{P_i} \, \partial C_i / \partial x$, and C_i is the comb-drive interelectrode capacitance. Next, the resonator stiffness, K_{ri}, is obtained. This is related to the capacitance, C_x, by

$$K_{i,res} = \frac{\eta_i^2}{C_x} \tag{6.20}$$

Next, the mechanical damping of the resonators, D, is obtained from

$$D_{i,res} = R_x \eta_i^2 \tag{6.21}$$

This concludes the specification for the mechanical design of the resonators. Next, the interresonator coupling springs are determined.

While the stiffness of simple beams may be used as springs, Nguyen [27] has pointed out that their nonzero mass can seriously degrade the performance of filters, because it can add to the effective mass of the resonators they couple, which modifies their resonance frequencies. To avoid this, he recommends two options. In one, he suggests the use of design methods that model coupling beams as impedances and that take advantage of the transmission line–like behavior of flexural-mode coupling beams at the filter center frequencies. When this method is used and the coupling beam lengths are chosen to be one-quarter wavelength at the filter center frequency, he argues, the beams appear to be massless to the adjacent resonators (microwave engineers will recognize this as a high-impedance transmission line stub). Johnson [29] gave the design equation for these beams as the simultaneous solution to the following two equations:

$$\sin \alpha \sinh \alpha + \cos \alpha \cosh \alpha = 0 \tag{6.22}$$

$$K_{spring,i,j} = \frac{-EI\alpha^3 (\sin \alpha + \sinh \alpha)}{L_{i,j}^3 (\cos \alpha \cosh \alpha - 1)} \tag{6.23}$$

where for vertical operation, $\alpha = \left(\rho A \omega^2 / EI \right)^{1/4}$, $I = W_{ij} h^3 / 12$, and $K_{spring,i,j}$ is the spring stiffness of the spring coupling resonators i and j, given by [27]

$$K_{spring, \ i,j} = K_r K_{i,j} \tag{6.24}$$

where $K_r = \sqrt{K_{ri} \cdot K_{rj}}$, and K_{ri} is the spring constant value of resonator i. From this the beam lengths, L_{ij}, and widths, W_{ij}, are obtained.

In a second approach to coupling spring realization, Nguyen [27] proposes attaching them to the folding trusses, instead of to the central mass. This approach, he argues, results in a greater effective spring stiffness, because the coupling point moves at a lower velocity. In particular, his technique should simplify beam design, since it yields beam widths and lengths with more manu-facturable dimensional features.

The last step in the design of MEM-coupled resonator filters is to realize the input and output couplings. One way in which this is realized employs resistors, connected in series with the input and output ports, whose values are chosen as [27]:

$$R_{in} = \frac{1}{2}\left(\frac{Q_{res,in}}{Q_{x,in}} - 1\right)R_{x,in} \tag{6.25}$$

for the input port, and as:

$$R_{out} = \frac{1}{2}\left(\frac{Q_{res,out}}{Q_{x,out}} - 1\right)R_{x,out} \tag{6.26}$$

for the output port. Q_{res} is the "turn-on" value of the resonators, and Q_x is its value as derived from the electrical simulation.

The physical design parameters, configuration and performance of proto-typical MEM-resonator filters are given in Tables 6.3 and 6.4, and Figures 6.11 and 6.12.

Table 6.3
Two-Resonator Filter Parameters

Resonator beam length	150	μm
Resonator beam width	1.7	μm
Resonator beam thickness	1.8	μm
Spring length	150	μm
Spring width	0.8	μm
Center frequency (f_0)	18.7	kHz
3-dB bandwidth (BW_3)	1.2	kHz
20-dB bandwidth (BW_{20})	3.2	kHz
Fractional bandwidth	6.4	%
$\left(\dfrac{BW_3}{f_0}\right)$		
Passband ripple	1.5	dB

Table 6.3 (continued)

Shape factor $\left(\dfrac{BW_{20}}{BW_3}\right)$	2.7	—
Quality factor $\left(\dfrac{f_0}{BW_3}\right)$	15.6	—
Dimensions	0.15	mm^2

Table 6.4
Three-Resonator Filter Parameters

Folded-beam length, L_r	32.8	μm
Folded-beam width, W_r	2	μm
Structural layer thickness	2	μm
Resonator effective mass, m_r	1.52×10^{-10}	Kg
Resonator spring constant, k_r	135.4	N/m
Comb-finger gap spacing, d	1	μm
Comb-finger gap overlap, L_0	5	μm
Electromechanical coupling,	1.48×10^{-7}	VF/m
Coupling beam length, $L_{12} = L_{23}$	75.2	μm
Coupling beam width, $W_{12} = W_{23}$	0.8	μm
Center frequency (f_0)	299.2	kHz
3-dB bandwidth (BW_3)	510	Hz
Quality factor	590	—
Stopband rejection	38	dB
Shape factor $\left(\dfrac{BW_{20}}{BW_3}\right)$	1.45	—
Insertion loss	<3	dB
Young's modulus, E	150	GPa
Density of polysilicon,	2,300	Kg/m^3
Filter dc bias, V_P	150	V

While the performance of integrated MEM-resonator-based filters is encouraging, much work still needs to be done in order to bring it to maturity. In particular, progress in the areas of CAD, linking filter fabrication and electrical performance, and the development of clever design techniques, will be critical to reduce design iterations. Further, to fully exploit the potentially high Qs of MEM resonators, it will be necessary to operate the filters in the context of a vacuum environment Thus packaging must be part of the design paradigm.

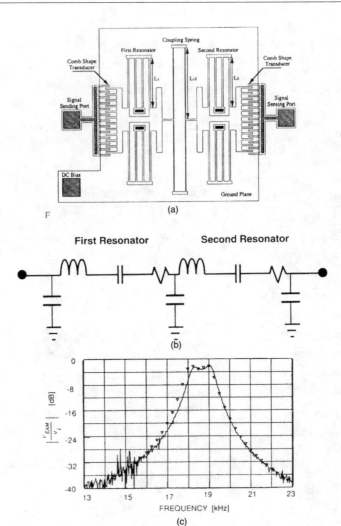

Figure 6.11 Two-resonator filter. (a) Schematic diagram. (*From*: [26] © 1992 IEEE. Reprinted with permission.) (b) Equivalent electrical circuit model. (c) Typical response diagram. (*From*: [26] © 1992 IEEE. Reprinted with permission.)

6.3.3.3   FBAR Filters

Film bulk acoustic wave resonators (FBAR)-based filters exploit the high-Q, high resonance frequency, and monolithic construction of FBAR resonators to enable monolithic filters at frequencies surrounding 1 GHz [31, 32]. Typical applications for which they are being currently produced and developed include 900-MHz cell phone diplexers, and 1.9-GHz filters for replacing mature discrete technologies employing resonators based on lumped element, ceramic, and surface acoustic wave (SAW) resonators. In this context, what is being sought

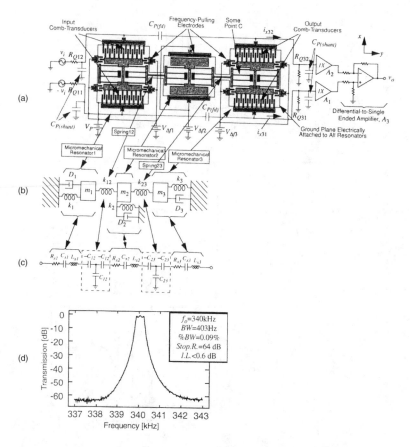

Figure 6.12 Three-resonator filter: (a) Schematic diagram,(b) equivalent lumped parameter mechanical circuit, (c) equivalent electrical circuit model, and (d) typical response. (*From*: [27], © 1997 IEEE. Reprinted with permission.)

are solutions that address the issues of on-chip integration with silicon RFICs, large power handling capability, low insertion loss, and small size and weight.

The most common topology employed to realize FBAR filters is the ladder one, as shown in Figure 6.13. In this topology, the series branches are populated with resonators possessing nominally identical series resonance frequencies (f_s) coinciding with the band center; while the shunt branches are populated by resonators possessing nominally identical series resonance frequencies (f_p) coinciding with the lower edge of the band. The upper edge of the band is set by the parallel resonance of the series branch resonators. The design of FBAR filters, as shown in Figure 6.13(b), consists in adjusting the resonators' f_s and f_p to shape the desired frequency response. The frequency response, in turn, is set by the bandwidth, which is given by: (1) the difference between the parallel resonance of the series resonators and the series resonance of the shunt resonators; (2) the

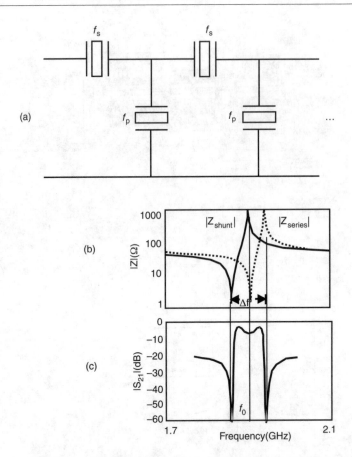

Figure 6.13 (a) FBAR filter topology. (b) relationship between location of series and branch resonator resonances and shaping of passband, and (c) passband corresponding to (b).

insertion loss which, as suggested in Figure 6.13(a), is determined by the minimum series resistance of the series resonators and the maximum shunt resistance of the parallel resonators; and (3) the out-of-band rejection, which is determined by the voltage divider ratio of the series and parallel resonators [33].

A 1.9 GHz seven-resonator FBAR filter, based on the membrane-supported resonators with AlN piezoelectric and molybdenum electrodes, was recently demonstrated by Park, et al. [33]. The measured performance of the FBAR filter was as follows: center frequency, 1.95 GHz; passband, 1.92 to 1.98 GHz; insertion loss of 2.8 dB; and passband ripple (PBR), 0.5 dB. These I.L. and PBR compared favorably with those exhibited by a commercial SAW filter, namely, 3.2 dB and 1.2 dB, respectively. A photo of a seven-resonator filter, and its corresponding measured performance, is shown in Figure 6.14(a, b).

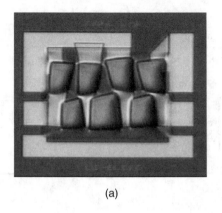

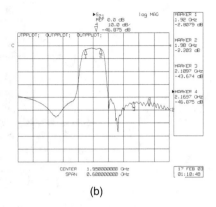

(a)   (b)

Figure 6.14 (a) Photo of seven-resonator FBAR filter, size 3 × 3 mm, and (b) plot of transmission response (S_{21}).

6.3.4 Oscillators

6.3.4.1 Oscillator Fundamentals

At the most fundamental level, an electric oscillator is a circuit which, with no input signal other than a dc bias, produces an ac signal of a certain amplitude and frequency. To see how this can come about, again refer to the concept of a resonator as a frequency memory element in the context of a single resonator filter, as shown in Figure 6.10(c), and let us consider the following three scenarios. First, suppose that the resonator energy has reached a steady state as a result of the excitation provided by the input signal S_i^+, but that all of a sudden S_i^+ is made zero (i.e., removed). What happens to the resonator energy? Since no energy is replenishing that which is being dissipated in the internal processes, as well as escaping through both the input and output ports, the resonator energy will immediately begin to decrease until it is reduced to zero. Next, suppose that, instead of allowing the energy to escape through the input and output ports, the power reaching the output is coupled to no other place except the input. What happens to the resonator energy? Again, while it may take longer than in the previous case for all of the stored resonator energy to be dissipated, it ultimately will be also reduced to zero. Finally, consider the case in which, as above, the output power is coupled to the input, but this time through an amplifier of power gain A, Figure 6.15. What happens to the resonator energy?

In this case, if the amplifier gain can make up for the energy dissipated in the resonator, then the stored energy will be sustained indefinitely, provided the following well-known conditions are met. First, in a trip around the loop the signal picks up a phase shift of 180, and second, the amplifier gain, A, is chosen so that it obeys

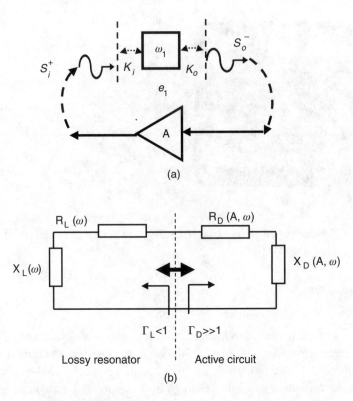

Figure 6.15 (a) Transmission resonator with feedback, and (b) negative resistance oscillator topology.

$$A \geq \frac{1}{K_i^2 K_o^2 \tau^2} \tag{6.27}$$

When the product of the amplifier and filter gains, $A \cdot A_f = -1$ [35], this latter arrangement behaves as an oscillator, with oscillation frequency ideally equal to that of the resonator. As long as the loop gain is greater than unity, the oscillation amplitude continues to increase, which, obviously, cannot go on forever in a physical system. In practice, the maximum steady state amplitude of the oscillation will settle to a value determined typically by gain saturation in the amplifier; that is, at large signal amplitudes the amplifier will be driven into its nonlinear region of operation. As in this region, its gain falls below unity, and the system will start losing energy. This continues until the amplifier is again in its linear region, where it resumes amplifying. The steady state amplitude reflects these dynamics.

At microwave- and millimeter-wave frequencies, another conceptual model of an oscillator is usually employed [36–39]. This is the negative resistance oscillator concept, as shown in Figure 6.15(b). The fundamental idea is that the power of a wave bouncing back and forth in a cavity formed by a frequency-determining circuit, possessing a reflection coefficient $\Gamma_L < 1$, and an active circuit possessing a reflection coefficient $\Gamma_D \gg 1$, but with round-trip product $\Gamma_L \cdot \Gamma_D > 1$, will increase until a nonlinear limiting mechanism effects saturation. At this point, $\Gamma_L \cdot \Gamma_D = 1$. The frequency-determining circuit and the active circuits may each be realized by a variety of means such as an LC-, cavity-, transmission line-, or MEM-resonator, in the former case; and a bipolar junction transistor, field-effect transistor, or a resonant-tunneling diode, in the latter. Physically, the power losses in the resonator, which occur due to its finite Q, are compensated for by the negative resistance of the active circuit.

The performance of oscillator circuits is characterized by the power of the output signal they produce, and by the stability of their output frequency. Since oscillator frequency stability is mathematically defined as [35]

$$S_f = \frac{d\theta}{d\omega}\bigg|_{\omega_0} \tag{6.28}$$

which for a parallel LCR resonator circuit is $S_f = -2Q$ (i.e., is proportional to the quality factor), it is clear that the higher the resonator Q, the more stable the frequency.

The recent progress in the development of FBAR resonators and their great potential for on-chip integration has drawn attention to their exploitation in applications usually reserved for SAW-based oscillators. In particular, low-jitter clocks in wired and wireless applications in the 500-MHz to- 5-GHz frequency range, which is currently beyond the capability of MEM resonators, have been identified as a potential niche for FBAR-based oscillators [40]. Specific advantages, with respect to SAW-based oscillators, included smaller size and improved frequency stability, which, in turn, are expected to endow FBAR oscillators with small size, high performance, and low cost (due to integration). In the next section, an FBAR-based negative resistance oscillator is described.

6.3.4.2 FBAR Voltage-Controlled Oscillator Circuits

The FBAR VCO, as shown in Figure 6.16, demonstrated by Khanna, Gane, and Chong [40], was of the negative resistance type, the frequency-determining element being embodied by an FBAR with 2 GHz *series* resonance frequency and an unloaded Q of 500. The active circuit topology was configured around a common-collector bipolar transistor, where the negative resistance was seen at its base, and the limiting action was produced by its emitter. A hyperabrupt junction varactor was for tuning purposes, and the assembly was of the "chip-and-wire" type, in which interconnection microstrip lines were printed on a

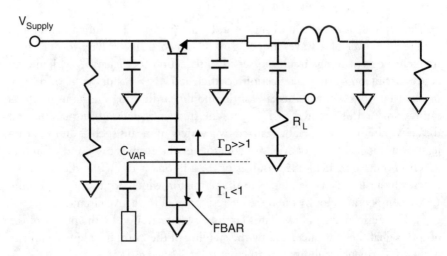

Figure 6.16 FBAR-based resonator voltage-controlled oscillaror. (*After:* [40].)

10-mil-thick, 250-×250-mil alumina substrate. The measured performance was as follows: frequency, 1,985 MHz; output power, 10 dBm; tuning sensitivity, 180 to 370 kHz/V; frequency drift over temperature, 0.4 MHz; frequency pushing (+/−5%), 200 kHz; second harmonic, −40 dBc; phase noise (roughly, ratio of power at a certain frequency away from the carrier, to carrier power, expressed in dB), −112 dBc/Hz @10 kHz; phase jitter over 12 kHz to 20 MHz, <0.1 ps; frequency tuning range, 2.5 MHz; tuning voltage range, <0 to 10V; and bias, 3.3V, 35 mA.

6.3.5 Packaging

6.3.5.1 Microwave Packaging Considerations

The proper operation of RF and microwave circuits and systems depends critically upon the "clean" environment provided by the package housing them. Indeed, packaging is considered an enabler for the commercialization of MEMS [41] for at least three reasons. First, due to the sensitive nature of their moving structures, MEMS must be protected against extraneous environmental influences (e.g., various forms of air contamination and moisture). Second, due to their small size, it is imperative that the devices be protected to withstand handling as they are integrated with other systems. Finally, since by their very nature RF and microwave circuits and systems are susceptible to electromagnetic coupling and moding, they must be electrically isolated. Moding refers to the resonant cavity-like behavior of metal structures enclosing high-frequency circuits, which can trap the energy being processed by these, and thus contribute an effective transmission loss extrinsic to the circuit. Two packaging concepts, addressing the environmental and electrical packaging aspects, have been advanced, namely, microriveting [41], and self-packaging [42].

Microriveting [41] is a wafer bonding technique aimed at performing *microscopic* device-level packaging, in the context of an IC process, so that the resulting integrated MEMS/IC wafer is ready to undergo the usual *macroscopic* packaging to which IC wafers are normally subjected. The technique requires a cap wafer, etched in a complementary way to the structures to be protected, so that when capped the structures find themselves within a cavity, and a seed metal layer for electroplating the rivets.

6.3.5.2 Wafer-Level Packaging

One of the attractive aspects of IC technology is its enablement of economies of scale in production, due to its ability to manufacture many devices simultaneously. This feature was extended to RF MEMS with the advent of packaging at the wafer level [i.e., wafer-level packaging (WLP)]. A number of techniques for WLP are available [43]. The essence of WLP, however, is exemplified by Tilmans, De Raedt, and Beyne, as shown in Figure 6.17 [43].

The approach, denominated "zero-level" packaging, consists of the utilization of a recessed wafer that is flipped and employed to cap the RF MEMS device. It is amenable, in particular, to package devices fabricated on the top layers of ICs and employs a bonding technique, preferably at temperatures low enough (~below 400°C) as to not perturb doping profiles and MEMS

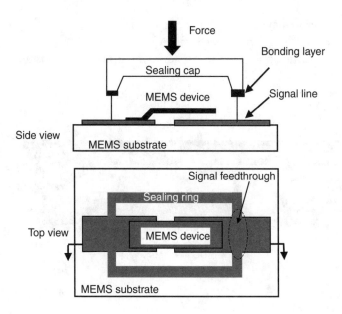

Figure 6.17 Sketch of wafer-level packaging. (*After:*[43].)

characteristics. One of the most important aspects of RF MEMS packaging is the need to prevent the package from introducing loss to the intrinsic RF MEMS characteristics. This necessitates careful attention in the design of transitions (i.e., the means by which the signals are brought in and out of the package). This technique is the hallmark of the self-packaging approach, aimed at microwave and millimeter-wave frequency applications.

6.3.5.3 Self-Packaging

Self-packaging [42, 44] is a technique aimed at both electrical and environmental protection of RF and microwave circuits. The fundamental idea is to make the package an integral part of the circuits being designed, rather than an afterthought. Thus, for example, the cavities formed are engineered to exhibit resonance frequencies well above the desired frequency band of operation. The technique also employs a second wafer, in which the complementary cavities and channels are etched, and which then caps the wafer containing the circuitry being packaged. Instances of the self-packaging approach have been described by Drayton and Katehi [42], and further developed by Margomenos, et al. [44], as shown in Figure 6.18, who have reported attaining packages displaying an insertion loss of only 0.1 dB and a return loss of 32 dB at 20 GHz.

6.4 Summary

In this chapter we have addressed the impact that MEMS technology can potentially exert on RF and microwave circuits and systems, particularly, in those functions that find application in wireless communications consumer products,

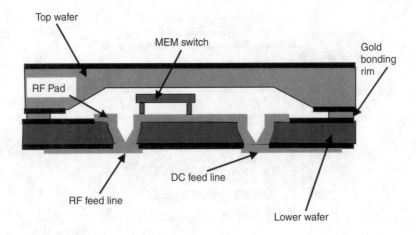

Figure 6.18 Sketch of RF MEMS self-packaging. (*After:* [44].)

but also in the infrastructure (base stations) and space-based communications systems. In general, MEMS technology is poised to be an enabler of low-insertion loss, passive, distributed components, monolithic active filters, and oscillators, ultralow-loss phase shifters, and large power-efficient phased-array antennas. In addition, the fact that a number of MEM-compatible packaging approaches are technically possible makes it very attractive for low-cost mass production ventures.

References

[1] Heilmeier, G. H., "Personal Communications: Quo Vadis," *IEEE Int. Solid-State Circuits Conf. Dig.*, February 1992, pp. 24–26.

[2] Abidi, A. A., "Low-Power Radio-Frequency ICs for Portable Communications," *Proc. IEEE*, 1995, pp. 544–569.

[3] De Los Santos, H. J., "MEMS—A Wireless Vision," *2001 International MEMS Workshop*, Singapore, July 4–6, 2001.

[4] Comtois, J. H., "Structures and Techniques for Implementing and Packaging Complex, Large Scale Microelectromechanical Systems Using Foundry Fabrication," Ph.D. dissertation, Air Force Institute of Technology, 1996.

[5] Maral, G., and M. Bousquet, *Satellite Communications Systems: Systems, Techniques and Technology*, 2nd Edition, New York: John Wiley & Sons, Inc., 1993.

[6] De Los Santos, et al., "MEMS-Based Communications Systems for Space-Based Applications," *Proc. Int. Conf. Micro/Nanotechnology for Space Appl.*, Houston, TX, October 30–November 3, 1995.

[7] Fischer, G., W. Eckl, and G. Kaminski, "RF-MEMS and SiC/GaN as Enabling Technologies for a Reconfigurable Multiband/Multistandard Radio," *Bell Labs Technical Journal*, Vol. 7, No. 3, 2003, pp. 169–189.

[8] Kaiser, A., "The potential of RF MEMS components for reconfigurable RF interfaces in mobile communications terminals,"Available [on-line]: http://www.imec.be/esscirc/esscirc2001/C01_Presentations/408.pdf

[9] Sokolov, V., et al., "A GaAs Monolithic Phase Shifter for 30 GHz Applications," *IEEE Microwave and Millimeter Wave Monolithic Circuits Symposium*, 1983, Boston, MA: pp. 40-44.

[10] Mailloux, R. J., "Phased Array Antenna and Technology," *Proc. IEEE*, Vol. 70, No. 3, March 1982, pp. 246–291.

[11] Koul, S., and Bhat, B., *Microwave and Millimeter Wave Phase Shifters*, Volume II, Norwood, MA: Artech House, Inc, 1991.

[12] De Los Santos, H. J., *RF MEMS Circuit Design for Wireless Communications*, Norwood, MA: Artech House, June, 2002.

[13] De Los Santos, H. J., et al., "Microwave MEMS and Micromachining," in *Handbook of Microwave Components*, Kai Chang, (Ed.), New York: John Wiley & Sons, Inc., October 2003.

[14] Ramo, S., *Introduction to Microwaves*, New York: McGraw-Hill, 1945.

[15] Chang, K., *Handbook of Microwave and Optical Components*, Volume I, New York: John Wiley & Sons, Inc., 1989.

[16] Nguyen, C. T.-C., "High-Q Microelectromechanical Oscillators and Filters for Communications," *IEEE Int. Symp. Circuits and Systems*, Hong Kong, June 9–12, 1997, pp. 2825–2828.

[17] Wang, K., et al., "Frequency Trimming and Q-Factor Enhancement of Micromechanical Resonators Via Localized Filament Annealing," *Int. Conf. Solid-State Sensors and Actuators, Transducer '97*, Chicago, IL, June 16–19, 1997, pp. 109–112.

[18] Lin, L., et al., "Micro Electromechanical Filters for Signal Processing," *IEEE Mocroelectromechanical Systems Workshop,*1992, pp. 226–231.

[19] Ragan, G. L., Ed., *Microwave Transmission Circuits*, Vol. 9, M.I.T. Radiation Lab. Series, New York: McGraw-Hill, 1948, Ch. 10.

[20] Zverev, A. I., *Handbook of Filter Synthesis*, New York: John Wiley & Sons, Inc., 1967.

[21] Temes, G. C., and S. K. Mitra, *Modern Filter Theory and Design*, New York: John Wiley & Sons, Inc., 1973.

[22] Datta, S., *Surface Acoustic Wave Devices*, Englewood Cliffs, NJ: Prentice-Hall, 1986.

[23] Haus, H. A., *Waves and Fields in Optoelectronics*, Englewood Cliffs, NJ: Prentice-Hall, Inc., 1984.

[24] Maxfield, J. P., and H. C. Harrison, "Method of High Quality Recording and Reproducing of Music and Speech Based on Telephone Research," *Bell System Tech. J.*, Vol. 5, July 1926, pp. 493–523.

[25] Gregorian, R., and G. C. Temes, *Analog MOS Integrated Circuits for Signal Processing*, New York: John Wiley & Sons, Inc., 1986.

[26] Lin, L., et al., "Micro Electromechanical Filters for Signal Processing," *IEEE Microelectro Mechanical Systems Workshop,* 1992, Travermunde, Germany: pp. 226–231.

[27] Wang, K., and C. T.-C. Nguyen, "High-Order Micromechanical Electronic Filters," *IEEE Microelectro Mechanical Systems Workshop,* Nagoya, Japan: 1997, pp. 25–30.

[28] Clark, J. R.,et al., "Parallel-Resonator HF Micromechanical Bandpass Filters," *IEEE Int. Conf. on Solid-State Sensors and Actuators,* Chicago, IL, June 16–19, 1997, pp. 1161–1164.

[29] Johnson, R. A., *Mechanical Filters in Electronics*, New York: John Wiley & Sons, Inc., 1983.

[30] Wang, K., et al., "Q-Enhancement of Microelectromechanical Filters Via Low-Velocity Spring Coupling," *IEEE Ultrasonics Symposium*, Toronto, Ontario, Canada: 1997, pp. 323–327.

[31]   Greiner, K., et al., "Integrated RF MEMS for Single Chip Radio," *Transducers'01*, Germany, Munich, Germany, June 10–14, 2001.

[32]   Bradley, P., R. Ruby, and J. D. Larson III, "A Film Bulk Acoustic Resonator (FBAR) Duplexer for USPCS," *2001 IEEE Int. Microwave Symp.*, Phoenix, AZ, May 20–25.

[33]   Park, J. Y., et al., "Comparison of Micromachined FBAR Band Pass Filters with Different Structural Geometry," *2003 IEEE Int. Microwave Symp. Digest.* June 8–13.

[34]   Park, J. Y., et al., "Silicon Bulk Micromachined FBAR Filters for W-CDMA Applications," *33rd European Microwave Conference*, Munich, Germany, 2003, pp. 907–910.

[35]   Millman, J., and C. C. Halkias, *Integrated Electronics: Analog and Digital Circuits and Systems*, New York: McGraw-Hill, 1972.

[36]   Boyles, J. W., "The Oscillator as a Reflection Amplifier: An Intuitive Approach to Oscillator Design," *Microwave J.*, June 1986, pp. 83–98.

[37]   Dougherty, R. M., "MMIC Oscillator Design Techniques," *Microwave J.*, August 1989, pp. 161–162.

[38]   "Microwave Oscillator Design," *Hewlett-Packard* Application Note A008.

[39]   Kurokawa, K., "Some Basic Characteristics of Broadband Negative Resistance Oscillator Circuits," *Bell Syst. Tech. J.*, Vol. 48, July 1969, pp. 1937–1955.

[40]   Khanna, A. P. S., E. Gane, and T. Chong, "A 2 GHz Voltage Tunable FBAR Oscillator," *2003 IEEE MTT-S Digest*, June 8–13, pp. 717–720.

[41]   Shivkumar, B., and C.-J. Kim, "Microrivets for MEMS Packaging: Concept, Fabrication, and Strength Testing," *J. Microelectromechanical Syst.*, Vol. 6, No. 3, 1997, pp. 217–225.

[42]   Drayton, R. F., and L. P. B. Katehi, "Micromachined Conformal Packages for Microwave and Millimeter-Wave Applications," *IEEE MTT-S Digest*, 1995, pp. 1387–1390.

[43]   Tilmans, H. A. C., W. De Raedt, and E. Beyne, "MEMS for Wireless Communications: 'From RF-MEMS Components to RF-MEMS-SiP'," *J. Micromech. Microeng.* Vol. 13, 2003, pp. S139–S163.

[44]   Margomenos, A., et al., "Silicon Micromachined Packages for RF MEMS Switches," *2001 European Microwave Conf.* London, U.K., September 25–27, pp. 100–103.

Glossary

CAD	computer-aided design
CMOS	complementary metal oxide semiconductor
CPW	coplanar waveguide
C-V	capacitance-voltage characteristic
dB	decibel
DCS	dichlorosilane
DRIE	deep RIE
EDP	diamine-pyrocatecol water
FET	field-effect transistor
GHz	gigahertz
GSG	ground-signal-ground
IC	integrated circuit
kHz	kilohertz
KOH	potassium hydroxide
LIGA	German acronym, consisting of the letters LI (RoentgenLIthographie, meaning X-ray lithography); G (Galvanik, meaning electrodeposition); and A (Abformung, meaning molding) process.
LO	local oscillator

LPCVD	low pressure chemical vapor deposition
MEM	microelctromechanical
MHz	megahertz
MIM	metal-insulator-metal
MIMAC	micromachined microwave actuator
MMIC	monolithic microwave integrated circuit
MOCVD	metalorganic chemical vapor deposition
MUMPs	multiuser MEM processes
NMOS	n-type metal oxide semiconductor
PECVD	plasma enhanced chemical vapor deposition
PMMA	poly-methil methacrylate
PZT	lead-zirconate-titanate
RF	radio frequency
RIE	reactive ion etching
SLIGA	sacrificial LIGA
SMA	shape memory alloy
SOI	silicon-on-insulator
SSPA	solid-state power amplifier
TEM	transverse electromagnetic

About the Author

Héctor J. De Los Santos is President and CTO of NanoMEMS Research, LLC, Irvine, CA, a new company engaged in RF MEMS and Nanotechnology research, consulting, and education, with main activities aiming at discovering and exploiting fundamentally new devices and circuits enabled by these novel technologies. Prior to founding NMR in October 2002, he spent 2 years as Principal Scientist at Coventor, Inc., Irvine, CA, where he led Coventor's RF MEMS intellectual property R&D effort, with activities including the conception, modeling, and design of novel RF MEMS devices and circuits. He received a Ph.D. from the School of Electrical Engineering, Purdue University, West Lafayette, IN, in 1989. From March 1989 to September 2000, he was employed at Hughes Space and Communications Company, Los Angeles, CA, where he served as Scientist and Principal Investigator and Director of the Future Enabling Technologies IR&D Program. Under this program, he pursued research in the areas of RF MEMS, Quantum Functional Devices and Circuits, and Photonic Bandgap Devices and Circuits. Dr. De Los Santos holds 18 U.S. and European patents, and is the author of the bestseller textbooks *Introduction to Microelectromechanical (MEM) Microwave Systems,* Norwood, MA: Artech House, 1999; and *RF MEMS Circuit Design for Wireless Communications,* Norwood, MA: Artech House, June 2002. Dr. De Los Santos' achievements are recognized in Marquis' *Who's Who in Science and Engineering,* Millenium Edition, and *Who's Who in the World,* 18th Edition. He is a Senior Member of the IEEE, and member of Tau Beta Pi, Eta Kappa Nu, and Sigma Xi. He is an IEEE Distinguished Lecturer of the Microwave Theory and Techniques Society for the 2001–2003 term.

Index

Recent Titles in the Artech House Microelectromechanical Systems (MEMS) Series

MEMS Mechanical Sensors, Steven Beeby, Graham Ensell, Michael Kraft, and Neil White

Fundamentals and Applications of Microfluidics, Nam-Trung Nguyen and Steven T. Wereley

Introduction to Microelectromechanical (MEM) Microwave Systems, Second Edition, Héctor J. De Los Santos

An Introduction to Microelectromechanical Systems Engineering, Nadim Maluf

MEMS Mechanical Sensors, Stephen Beeby et al.

RF MEMS Circuit Design for Wireless Communications, Héctor J. De Los Santos

For further information on these and other Artech House titles, including previously considered out-of-print books now available through our In-Print-Forever® (IPF®) program, contact:

Artech House
685 Canton Street
Norwood, MA 02062
Phone: 781-769-9750
Fax: 781-769-6334
e-mail: artech@artechhouse.com

Artech House
46 Gillingham Street
London SW1V 1AH UK
Phone: +44 (0)20 7596-8750
Fax: +44 (0)20 7630-0166
e-mail: artech-uk@artechhouse.com

Find us on the World Wide Web at:
www.artechhouse.com